普通高等教育"十三五"规划教材

English for Professional Engineers

Materials Forming and Control Engineering

(Second Edition)

Xu Guang　　Hu Haijiang
Pan Chenggang　Yan Wenqing

北　京
冶金工业出版社
2024

内 容 提 要

本教材为高等院校材料成型与控制工程专业的专业英语教学用书，全书共 15 章，第 1 章介绍材料科学相关基础知识；第 2 章介绍金属材料强化机理等内容；第 3~7 章为轧制方面的内容，包括轧制的发展历史、轧机的分类、中厚板轧机、热带轧机以及最新的薄带双辊铸造技术等；第 8 章为铸造方面的内容，主要介绍金属铸造工艺；第 9~11 章为锻造、冲压和挤压方面的内容，包括金属材料锻造分类和设备、冲压成型方法、挤压分类及原理等；第 12~15 章为焊接方面的内容，包括焊接过程的分类、焊接原理、手工电弧焊技术及应用和气体保护焊技术及应用等；附录为钢铁材料领域学术论文实例。

本书既可作为高等院校材料成型与控制工程专业的专业英语教学用书，也可作为从事金属材料研究与生产的科技人员的参考书。

图书在版编目（CIP）数据

专业英语教程：材料成型与控制工程＝English for Professional Engineers—Materials Forming and Control Engineering（Second Edition）：英文／徐光等编著 . —2 版 . —北京：冶金工业出版社，2017.7（2024.8 重印）
普通高等教育"十三五"规划教材
ISBN 978-7-5024-7552-9

Ⅰ.①专⋯ Ⅱ.①徐⋯ Ⅲ.①金属加工—高等学校—教材—英文 Ⅳ.①TG

中国版本图书馆 CIP 数据核字（2017）第 180249 号

English for Professional Engineers
Materials Forming and Control Engineering（Second Edition）

出版发行	冶金工业出版社	电　　话	（010）64027926
地　　址	北京市东城区嵩祝院北巷 39 号	邮　　编	100009
网　　址	www.mip1953.com	电子信箱	service@ mip1953.com

责任编辑　曾　媛　美术编辑　吕欣童　版式设计　孙跃红
责任校对　李　娜　责任印制　禹　蕊

北京富资园科技发展有限公司印刷
2006 年 12 月第 1 版，2017 年 7 月第 2 版，2024 年 8 月第 6 次印刷
787mm×1092mm　1/16；11.5 印张；279 千字；174 页
定价 35.00 元

投稿电话　（010）64027932　投稿信箱　tougao@cnmip.com.cn
营销中心电话　（010）64044283
冶金工业出版社天猫旗舰店　yjgycbs.tmall.com
（本书如有印装质量问题，本社营销中心负责退换）

第 2 版前言

随着我国高等院校专业调整，原有的轧制、铸造、模具、焊接等专业合并为一个新的本科专业——材料成型与控制工程专业。为了适应合并后的材料成型与控制工程专业英语的教学要求，我们编写了《English for Professional Engineers—Materials Forming and Control Engineering》。该书自出版以来，已被很多高校材料成型与控制工程专业教学采用。为了适应形势的发展和专业要求，作者对本书进行了修订，扩充了材料科学领域相关基础知识，其中包括晶格结构、晶系、材料强化和软化机制等，专业英语知识涉及面更为广泛、更为系统。附录给出了作者发表的几篇学术论文，供本专业研究生和工程技术人员撰写英文学术论文参考，也可以作为研究生专业文献阅读资料。

本书第 1 和 2 章、第 6~8 章由胡海江修订，第 3~5 章由徐光修订，第 9~11 章由潘成刚修订，第 12~15 章由闫文青修订，全书由徐光审定定稿。

本教材根据作者多年专业英语教学实践和经验编写而成，读者对象为高等院校材料成型与控制工程专业专科、本科生和研究生，以及本领域内工程技术与科研人员。

由于时间仓促，加之作者水平所限，书中不妥之处，恳请读者批评指正。

徐 光
2017 年 5 月 1 日

第1版前言

随着我国高等院校专业调整，原有的轧制、铸造、模具、焊接等专业合并为一个新的本科专业——材料成型与控制工程专业。原有的上述各专业专业英语教材已不适应合并后专业的英语教学需求。为了适应合并后的材料成型与控制工程专业英语教学要求，我们编写了本教材。

本教材由武汉科技大学材料成型与控制工程系徐光教授、张诗昌副教授、潘成刚和闫文青讲师共同编著。本书共分16章，其中第1章~第6章由徐光编写，主要涉及轧制方面的内容；第7章~第9章由张诗昌编写，主要涉及铸造方面的内容；第10章~第12章由潘成刚编写，主要涉及其他金属塑性成型和模具方面的内容；第13章~第16章由闫文青编写，主要涉及焊接方面的内容。全书由徐光审定。

本教材是作者根据多年专业英语教学实践和经验编写而成，读者对象为高等院校材料成型与控制工程专业专科、本科生，以及研究生和本领域内工程技术人员。

由于作者水平所限，书中不妥之处在所难免，恳请专家和读者批评指正。

徐 光
2006年11月2日

Content

1 Basic Concepts of Metallic Materials ... 1

- 1.1 Introduction ... 1
- 1.2 Metallic Crystals Structures ... 1
  - 1.2.1 The Face-centered Cubic Crystal Structure ... 2
  - 1.2.2 The Body-centered Cubic Crystal Structure ... 3
  - 1.2.3 The Hexagonal Close-packed Crystal Structure ... 4
- 1.3 Crystal Systems ... 5
- 1.4 Crystallographic Directions ... 7
- 1.5 Directions in Hexagonal Crystals ... 8
- 1.6 Crystallographic Planes ... 10
- Words and Expressions ... 12

2 Strengthening Mechanisms ... 13

- 2.1 Mechanisms of Strengthening in Metals ... 13
  - 2.1.1 Strengthening by Grain Size Reduction ... 13
  - 2.1.2 Solid-solution Strengthening ... 15
  - 2.1.3 Strain Hardening ... 17
- 2.2 Recovery, Recrystallization, and Grain Growth ... 18
  - 2.2.1 Recovery ... 19
  - 2.2.2 Recrystallization ... 19
  - 2.2.3 Grain Growth ... 22
- Words and Expressions ... 24

3 The History of Rolling ... 25

- 3.1 The Early History of Rolling ... 25
- 3.2 Late Developments in the Hot Rolling of Steel in Europe ... 27
- 3.3 The Early Rolling of Steel in the United States ... 29
- 3.4 Energy Source for Rolling Mills ... 31
- 3.5 The Historical Development of Cold Rolling ... 32
- 3.6 Modern Cold-reduction Facilities ... 33
- 3.7 Foil Mills ... 35

3.8 Temper or Skin Pass Mills ································ 35
3.9 Modern Hot Strip Mills ································ 35
Words and Expressions ································ 38

4 Classifications of Rolling Mills ································ 43

4.1 Main Components of a Mill Stand ································ 43
4.2 Classification of Mill Stands ································ 44
4.3 General Classification of Rolling Mills ································ 47
4.4 Components of High-production Hot Strip Mills ································ 48
4.5 Classification of High-production Hot Strip Mills ································ 49
4.6 Compact Hot Strip Mills ································ 50
4.7 Integrated Continuous Casting and Hot Rolling Process ································ 51
4.8 Cold Mill Arrangement ································ 52
4.9 Processing Lines Incorporating Cold Mills ································ 52
Words and Expressions ································ 53

5 Plate Mills ································ 56

5.1 Introduction ································ 56
5.2 Types of Mills Used for the Rolling of Plates ································ 56
5.3 Plate-mill Design ································ 57
Words and Expressions ································ 58

6 Hot-Strip Mills ································ 60

6.1 Introduction ································ 60
6.2 Steckel Hot Mills ································ 60
6.3 Planetary Mills ································ 61
6.4 Semi-continuous Hot-strip Mills ································ 62
6.5 Continuous Hot-strip Mills ································ 63
6.6 Rougher Trains ································ 65
6.7 Finishing Trains ································ 66
Words and Expressions ································ 66

7 Twin-roll Casting Technology ································ 68

7.1 Introduction ································ 68
7.2 Process Overview ································ 69
7.3 Nucor's Crawfordsville CASTRIP Facility ································ 71
7.4 The Limits of Present Thick and Thin Slab Casting ································ 73
Words and Expressions ································ 75

8 Metal Casting Processes ... 77

- 8.1 History ... 77
- 8.2 Advantages and Limitations ... 78
- 8.3 Applications ... 78
- 8.4 Casting Terms ... 78
- 8.5 Sand Mould Making Procedure ... 79
- 8.6 Melting Equipment for Non-ferrous Foundries ... 80
- 8.7 Fluidity and Pouring ... 82
- Words and Expressions ... 83

9 Forging ... 85

- 9.1 Classification of Forging Process ... 85
- 9.2 Forging Equipment ... 88
- 9.3 Open-die Forging ... 88
- 9.4 Close-die Forging ... 91
- 9.5 Calculation of Forging Loads in Closed-die Forging ... 93
- 9.6 Forging Defects ... 93
- 9.7 Residual Stresses in Forging ... 94
- Words and Expressions ... 95

10 Sheet-Metal Forming ... 97

- 10.1 Introduction ... 97
- 10.2 Forming Methods ... 98
- 10.3 Shearing and Blanking ... 102
- 10.4 Bending ... 103
- 10.5 Stretch Forming ... 104
- 10.6 Deep Drawing ... 106
- Words and Expressions ... 107

11 Extrusion ... 110

- 11.1 Classification of Extrusion Processes ... 110
- 11.2 Extrusion Equipment ... 111
- 11.3 Hot Extrusion ... 113
- 11.4 Cold Extrusion and Cold Forming ... 115
- 11.5 Hydrostatic Extrusion ... 116
- 11.6 Extrusion of Tubing ... 117
- Words and Expressions ... 118

12　Classification of Welding Processes ……… 120

Words and Expressions ……… 122

13　Methods of Welding ……… 124

13.1　Types of Welded Joint ……… 124
13.2　Weld Formation ……… 126
13.3　Cold Pressure Welding ……… 127
13.4　Hot Pressure Welding ……… 128
13.5　Gas Heating ……… 128
13.6　Resistance Heating ……… 128
13.7　Induction Heating ……… 129
13.8　Friction Welding ……… 130
Words and Expressions ……… 132

14　Welding Techniques for Manual Welding ……… 133

14.1　Operating Characteristics of Manual Metal Arc (MMA) Welding ……… 134
　　14.1.1　Welding Current ……… 134
　　14.1.2　Position of Welding ……… 136
　　14.1.3　Functions of Flux Covering ……… 138
14.2　Weld Metal Protection ……… 138
14.3　Arc Stabilization ……… 139
14.4　Control of Surface Profile ……… 139
14.5　Control of Weld Metal in Position ……… 140
14.6　Control of Weld Metal Composition ……… 140
Words and Expressions ……… 141

15　Gas Shielded Arc Welding ……… 142

15.1　Historical Background ……… 142
15.2　Inert Gas Tungsten Arc Welding ……… 144
15.3　Electrode Polarity ……… 145
15.4　Arc Maintenance ……… 146
15.5　Direct Current Component ……… 148
15.6　Starting the Welding Arc ……… 148
15.7　Welding Techniques ……… 149
15.8　Stopping the Weld ……… 150
15.9　Applications ……… 150
Words and Expressions ……… 151

Appendix 1: Academic Paper Ⅰ 152

Appendix 2: Academic Paper Ⅱ 156

Appendix 3: Academic Paper Ⅲ 162

Appendix 4: Academic Paper Ⅳ 167

References 174

1 Basic Concepts of Metallic Materials

1.1 Introduction

The present part is devoted to the *microstructure*❶ of materials, specifically, to some of the arrangements that may be assumed by atoms in the solid state. Within this framework, the basic concepts of *metallic crystal* are introduced. For metallic crystalline solids, the notion of crystal structure is presented, specified in terms of a unit cell. The three common crystal structures found in metals are then detailed, along with the scheme by which *crystallographic* directions and planes are expressed.

1.2 Metallic Crystals Structures

The atomic bonding in this group of materials is metallic. Consequently, there are minimal restrictions as to the number and position of nearest-neighbor atoms; this leads to relatively large numbers of nearest neighbors and dense atomic packings for most metallic crystal structures. Also, for metals, when we use the *hard-sphere model* for the crystal structure, each sphere represents an ion core. Table 1-1 presents the atomic *radii* for a number of metals. Three relatively simple crystal structures are found for most of the common metals: *face-centered cubic*, *body-centered cubic*, and *hexagonal close-packed*.

Table 1-1 Atomic radii and crystal structures for 16 metals

Metal	Crystal Structure①	Atomic Radius②/nm	Metal	Crystal Structure①	Atomic Radius②/nm
Aluminum	FCC	0.1431	Molybdenum	BCC	0.1363
Cadmium	HCP	0.1490	Nickel	FCC	0.1246
Chromium	BCC	0.1249	Platinum	FCC	0.1387
Cobalt	HCP	0.1253	Silver	FCC	0.1445
Copper	FCC	0.1278	Tantalum	BCC	0.1430
Gold	FCC	0.1442	Titanium(α)	HCP	0.1445
Iron(α)	BCC	0.1241	Tungsten	BCC	0.1371
Lead	FCC	0.1750	Zinc	HCP	0.1332

① FCC = face-centered cubic; HCP = hexagonal close-packed; BCC = body-centered cubic.
② A nanometer (nm) equals 10^{-9} m; to convert from nanometers to angstrom units (Å), multiply the nanometer value by 10.

---

❶ The italic technical terms are translated into Chinese at the end of chapters.

1.2.1 The Face-centered Cubic Crystal Structure

The crystal structure found for many metals has a *unit cell* of cubic geometry, with atoms located at each of the corners and the centers of all the cube faces. It is called the face-centered cubic (FCC) crystal structure. Fig. 1-1(a) shows a hard-sphere model for the FCC unit cell, whereas in Fig. 1-1(b) the atom centers are represented by small circles to provide a better perspective on atom positions. The *aggregate* of atoms in Fig. 1-1(c) represents a section of crystal consisting of many FCC unit cells. These spheres or ion cores touch one another across a face *diagonal*; the *cube edge length a* and the atomic radius R are related through.

$$a = 2R\sqrt{2} \qquad (1-1)$$

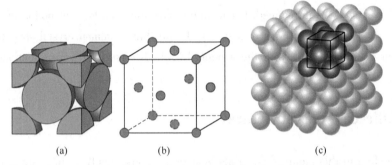

Fig. 1-1 For the face-centered cubic crystal structure, (a) a hard-sphere unit cell representation, (b) a reduced-sphere unit cell, and (c) an aggregate of many atoms

On occasion, we need to determine the number of atoms associated with each unit cell. Depending on an atom's location, it may be considered to be shared with adjacent unit cells, that is, only some fraction of the atom is assigned to a specific cell. For example, for cubic unit cells, an atom completely within the interior belongs to that unit cell, one at a cell face is shared with one other cell, and an atom located at a corner is shared among eight. The number of atoms per unit cell, N, can be calculated using the following formula:

$$N = N_i + \frac{N_f}{2} + \frac{N_c}{8} \qquad (1-2)$$

Where

N_i = the number of interior atoms

N_f = the number of face atoms

N_c = the number of corner atoms

For the FCC crystal structure, there are eight corner atoms ($N_c = 8$), six face atoms ($N_f = 6$), and no interior atoms ($N_i = 0$). Thus, from Eq. (1-2)

$$N = 0 + \frac{6}{2} + \frac{8}{8} = 4$$

so a total of four whole atoms may be assigned to a given unit cell. This can be discribed in Fig. 1-1(a), where only sphere portions are represented within the confines of the cube. The cell is composed of the volume of the cube that is generated from the centers of the corner atoms, as shown in the figure.

Corner and face positions are really equivalent, that is, change of the cube corner from an original corner atom to the center of a face atom will not alter the cell structure. Two other important characteristics of a crystal structure are the *coordination number* and the *atomic packing factor* (APF). For metals, each atom has the same number of nearest-neighbor or touching atoms, which is the coordination number. For face-centered cubics, the coordination number is 12. This may be confirmed by examination of Fig. 1-1(a); the front face atom has four corner nearest-neighbor atoms surrounding it, four face atoms that are in contact from behind, and four other equivalent face atoms located in the next unit cell to the front (not shown).

The APF is the sum of the sphere volumes of all atoms within a unit cell (assuming the atomic hard-sphere model) divided by the unit cell volume, that is,

$$\text{APF} = \frac{\text{volume of atoms in a unit cell}}{\text{total unit cell volume}} \tag{1-3}$$

For the FCC structure, the atomic packing factor is 0.74, which is the maximum packing possible for spheres all having the same diameter.

1.2.2 The Body-centered Cubic Crystal Structure

Another common metallic crystal structure also has a cubic unit cell with atoms located at all eight corners and a single atom at the cube center. This is called a body-centered cubic (BCC) crystal structure. A collection of spheres depicting this crystal structure is shown in Fig. 1-2(c), whereas Fig. 1-2(a) and Fig. 1-2(b) are diagrams of BCC unit cells with the atoms represented by hard-sphere and *reduced-sphere models*, respectively. Center and corner atoms touch one another along cube diagonals, and unit cell length a and atomic radius R are related by

$$a = \frac{4R}{\sqrt{3}} \tag{1-4}$$

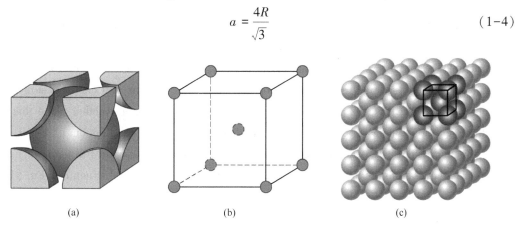

Fig. 1-2 For the body-centered cubic crystal structure, (a) a hard-sphere unit cell representation, (b) a reduced-sphere unit cell, and (c) an aggregate of many atoms

Chromium, iron, *tungsten*, and several other metals listed in Table 1-1 exhibit a BCC structure. Each BCC unit cell has eight corner atoms and a single center atom, which is contained within its cell; therefore, from Eq. (1-2), the number of atoms per BCC unit cell is

$$N = N_i + \frac{N_f}{2} + \frac{N_c}{8} = 1 + 0 + \frac{8}{8} = 2$$

The coordination number for the BCC crystal structure is 8; each center atom has as nearest neighbors its eight corner atoms. Because the coordination number is less for BCC than for FCC, the atomic packing factor is also lower for BCC—0.68 versus 0.74.

It is also possible to have a unit cell that consists of atoms situated only at the corners of a cube. This is called the simple cubic (SC) crystal structure; hard-sphere and reduced-sphere models are shown, respectively, in Fig. 1-3(a) and Fig. 1-3(b). None of the metallic elements have this crystal structure because of its relatively low atomic packing factor. The only simple-cubic element is *polonium*, which is considered to be a *metalloid* (or semi-metal).

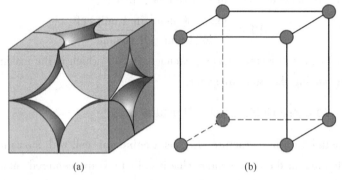

Fig. 1-3 For the simple cubic crystal structure, (a) a hard-sphere unit cell, and (b) a reduced-sphere unit cell

1.2.3 The Hexagonal Close-packed Crystal Structure

Not all metals have unit cells with cubic *symmetry*; the final common metallic crystal structure to be discussed has a unit cell that is hexagonal. Fig. 1-4(a) shows a reduced- sphere unit cell for this structure, which is termed hexagonal close-packed (HCP); an assemblage of several HCP unit cells is presented in Fig. 1-4(b). The top and bottom faces of the unit cell consist of six atoms that form regular hexagons and surround a single atom in the center. Another plane that provides three additional atoms to the unit cell is situated between the top and bottom planes. The atoms in this midplane have as nearest neighbors atoms in both of the adjacent two planes.

In order to calculate the number of atoms per unit cell for the HCP crystal structure, Eq. (1-2) is modified to read as follows:

$$N = N_i + \frac{N_f}{2} + \frac{N_c}{6} \tag{1-5}$$

That is, one-sixth of each corner atom is assigned to a unit cell (instead of 8 as with the cubic structure). Because for HCP there are 6 corner atoms in each of the top and bottom faces (for a total of 12 corner atoms), 2 face center atoms (one from each of the top and bottom faces), and

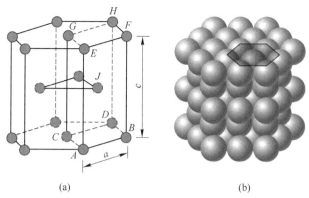

Fig. 1-4 For the hexagonal close-packed crystal structure, (a) a reduced-sphere unit cell (a and c represent the short and long edge lengths, respectively), and (b) an aggregate of many atoms

3 midplane interior atoms, the value of N for HCP is found, using Eq. (1-5), to be

$$N = 3 + \frac{2}{2} + \frac{12}{6} = 6$$

Thus, 6 atoms are assigned to each unit cell.

If a and c represent, respectively, the short and long unit cell dimensions of Fig. 1-4(a), the c/a ratio should be 1.633; however, for some HCP metals, this ratio deviates from the ideal value. The coordination number and the atomic packing factor for the HCP crystal structure are the same as for FCC: 12 and 0.74, respectively.

1.3 *Crystal Systems*

Because there are many different possible crystal structures, it is sometimes convenient to divide them into groups according to unit cell configurations and/or atomic arrangements. One such scheme is based on the unit cell geometry, that is, the shape of the appropriate unit cell parallelepiped without regard to the atomic positions in the cell. Within this framework, an *xyz* coordinate system is established with its origin at one of the unit cell corners; each of the x, y, and z axes coincides with one of the three parallelepiped edges that extend from this corner, as illustrated in Fig. 1-5. The unit cell geometry is completely defined in terms of six parameters: the three edge lengths a, b, and c, and the three *interaxial angles* α, β and γ. These are indicated in Fig. 1-5, and are sometimes termed the *lattice* parameters of a crystal structure.

On this basis, there are seven different possible combinations of a, b, and c and α, β and γ, each of which represents a distinct crystal system. These seven crystal systems are cubic, *tetragonal*, hexagonal, *orthorhombic*, *rhombohedral*, *monoclinic*, and *triclinic*. The lattice parameter relationships and unit cell sketches for each are represented in Table 1-2. The cubic system, for which $a=b=c$ and $\alpha=\beta=\gamma=90°$, has the greatest degree of symmetry. The least symmetry is displayed by the triclinic system, because $a \neq b \neq c$ and $\alpha \neq \beta \neq \gamma$.

From the discussion of metallic crystal structures, it is obvious that both FCC and BCC structures belong to the cubic crystal system, whereas HCP falls within the hexagonal system. The conventional hexagonal unit cell really consists of three parallelepipeds situated as shown in Table 1-2.

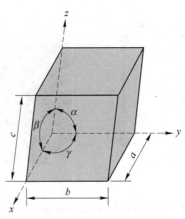

Fig. 1-5 A unit cell with x, y, and z coordinate axes, showing axial lengths (a, b, and c) and interaxial angles (α, β and γ)

Table 1-2 Lattice parameter relationships showing unit cell geometries for the seven crystal systems

Crystal System	Axial Relationships	Interaxial Angles	Unit Cell Geometry
Cubic	$a=b=c$	$\alpha=\beta=\gamma=90°$	
Hexagonal	$a=b\neq c$	$\alpha=\beta=90°$, $\gamma=120°$	
Tetragonal	$a=b\neq c$	$\alpha=\beta=\gamma=90°$	
Rhombohedral (Trigonal)	$a=b=c$	$\alpha=\beta=\gamma\neq 90°$	
Orthorhombic	$a\neq b\neq c$	$\alpha=\beta=\gamma=90°$	
Monoclinic	$a\neq b\neq c$	$\alpha=\gamma=90°\neq\beta$	
Triclinic	$a\neq b\neq c$	$\alpha\neq\beta\neq\gamma\neq 90°$	

1.4 Crystallographic Directions

A *crystallographic direction* is defined as a line directed between two points, or a vector. The following steps are used to determine the three directional *indices*:

(1) A right-handed x-y-z coordinate system is first constructed. As a matter of convenience, its origin may be located at a unit cell corner.

(2) The coordinates of two points that lie on the direction *vector* (referenced to the coordinate system) are determined, for example, for the vector tail, point 1: x_1, y_1, and z_1; whereas for the vector head, point 2: x_2, y_2, and z_2.

(3) Tail point coordinates are subtracted from head point components, that is, x_2-x_1, y_2-y_1, and z_2-z_1.

(4) These coordinate differences are then normalized in terms of (i.e., divided by) their respective a, b, and c lattice parameters—that is,

$$\frac{x_2 - x_1}{a} \quad \frac{y_2 - y_1}{b} \quad \frac{z_2 - z_1}{c}$$

which yields a set of three numbers.

(5) If necessary, these three numbers are multiplied or divided by a common factor to reduce them to the smallest *integer values*.

(6) The three resulting indices, not separated by commas, are enclosed in square brackets, thus: $[uvw]$. The u, v, and w integers correspond to the normalized coordinate differences referenced to the x, y, and z axes, respectively.

In summary, the u, v, and w indices may be determined using the following equations:

$$u = n\left(\frac{x_2 - x_1}{a}\right) \quad (1\text{-}6\,(a))$$

$$v = n\left(\frac{y_2 - y_1}{b}\right) \quad (1\text{-}6\,(b))$$

$$w = n\left(\frac{z_2 - z_1}{c}\right) \quad (1\text{-}6\,(c))$$

In these expressions, n is the factor that may be required to reduce u, v, and w to integers. For each of the three axes, there are both *positive and negative* coordinates. Thus, negative indices are also possible, which are represented by a bar over the appropriate index. For example, the $[1\bar{1}1]$ direction has a component in the $-y$ direction. Also, changing the signs of all indices produces an *antiparallel* direction; that is, $[1\bar{1}1]$ is directly opposite to $[\bar{1}1\bar{1}]$. If more than one direction (or plane) is to be specified for a particular crystal structure, it is important for maintaining consistency that a positive-negative convention, once established, not be changed.

The $[100]$, $[110]$, and $[111]$ directions are common ones; they are drawn in the unit cell shown in Fig. 1-6.

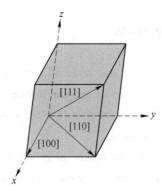

Fig. 1-6 The [100], [110], and [111] directions within a unit cell

For some crystal structures, several nonparallel directions with different indices are actually equivalent, meaning that the spacing of atoms along each direction is the same. For example, in cubic crystals, all the directions represented by the following indices are equivalent: [100], [$\bar{1}$00], [010], [0$\bar{1}$0], [001], and [00$\bar{1}$]. As a convenience, equivalent directions are grouped together into a family, which is enclosed in angle brackets, thus: $\langle 100 \rangle$. Furthermore, directions in cubic crystals having the same indices without regard to order or sign, for example, [123] and [$\bar{2}$1$\bar{3}$] —are equivalent. This is, in general, not true for other crystal systems. For example, for crystals of tetragonal symmetry, the [100] and [010] directions are equivalent, whereas the [100] and [001] are not.

1.5 Directions in Hexagonal Crystals

A problem arises for crystals having hexagonal symmetry in that some equivalent crystallographic directions do not have the same set of indices. For example, the [111] direction is equivalent to [$\bar{1}$01] rather than to a direction with indices that are combinations of 1s and −1s. This situation is addressed using a four-axis coordinate system, which is shown in Fig. 1-7. The three a_1, a_2, and a_3 axes are all contained within a single plane (called the *basal plane*) and are at 120° angles to one another. The z axis is *perpendicular* to this basal plane. Directional indices, which are obtained as described earlier, are denoted by four indices, as [$uvtw$]; by convention, the u, v, and t indices relate to vector coordinate differences referenced to the respective a_1, a_2, and a_3 axes in the basal plane; the fourth index pertains to the z axis.

Conversion from the three-index system to the four-index system as
$$[UVW] \rightarrow [uvtw]$$
is accomplished using the following formulas:

$$u = \frac{1}{3}(2U - V) \qquad (1-7(a))$$

$$v = \frac{1}{3}(2V - U) \qquad (1-7(b))$$

1.5 Directions in Hexagonal Crystals

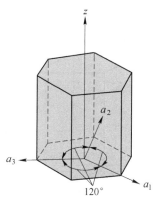

Fig. 1-7 Coordinate axis system for a hexagonal unit cell (Miller-Bravais scheme)

$$t = -(u + v) \qquad (1-7(c))$$
$$w = W \qquad (1-7(d))$$

Here, uppercase U, V, and W indices are associated with the three-index scheme (instead of u, v, and w as previously), whereas lowercase u, v, t, and w correlate with the four-index system. For example, using these equations, the [010] direction becomes [$\bar{1}2\bar{1}0$]. Several directions have been drawn in the hexagonal unit cell of Fig. 1-8.

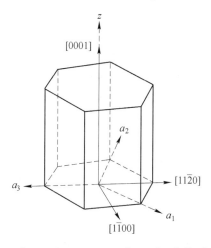

Fig. 1-8 For the hexagonal crystal system, the [0001], [$1\bar{1}00$] and [$11\bar{2}0$] directions

When plotting crystallographic directions for hexagonal crystals, it is sometimes more convenient to modify the four-axis coordinate system shown in Fig. 1-7 to that of Fig. 1-9; here, a grid has been constructed on the basal plane that consists of sets of lines parallel to each of the a_1, a_2, and a_3 axes. The intersections of two sets of parallel lines (e.g., those for a_2, and a_3) lie on and trisect the other axis (i.e., divide a_1 into thirds) within the hexagonal unit cell. In addition, the z axis of Fig. 1-9 is also apportioned into three equal lengths (at trisection points m and n). This scheme is sometimes referred to as a ruled-net coordinate system.

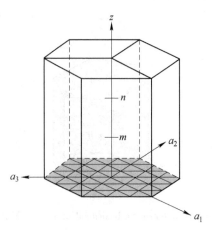

Fig. 1-9 Ruled-net coordinate axis system for hexagonal unit cells that may be used to plot crystallographic directions

1.6 Crystallographic Planes

The *orientations* of planes for a crystal structure are represented in a similar manner. Again, the unit cell is the basis, with the three-axis coordinate system as represented in Fig. 1-5. In all but the hexagonal crystal system, *crystallographic planes* are specified by three *Miller indices* as (hkl). Any two planes parallel to each other are equivalent and have identical indices. The procedure used to determine the h, k, and l index numbers is as follows:

(1) If the plane passes through the selected origin, either another parallel plane must be constructed within the unit cell by an appropriate translation, or a new origin must be established at the corner of another unit cell.

(2) At this point, the crystallographic plane either intersects or parallels each of the three axes. The coordinate for the intersection of the crystallographic plane with each of the axes is determined (referenced to the origin of the coordinate system). These *intercepts* for the x, y, and z axes will be designed by A, B, and C, respectively.

(3) The *reciprocals* of these numbers are taken. A plane that parallels an axis is considered to have an infinite intercept and therefore a zero index.

(4) The reciprocals of the intercepts are then normalized in terms of (i.e., multiplied by) their respective a, b, and c lattice parameters. That is,

$$\frac{a}{A} \quad \frac{b}{B} \quad \frac{c}{C}$$

(5) If necessary, these three numbers are changed to the set of smallest integers by multiplication or by division by a common factor.

(6) Finally, the integer indices, not separated by commas, are enclosed within parentheses, thus: (hkl). The h, k, and l integers correspond to the normalized intercept reciprocals referenced to the x, y, and z axes, respectively.

In summary, the h, k, and l indices may be determined using the following equations:

$$h = \frac{na}{A} \quad (1\text{-}8(\text{a}))$$

$$k = \frac{nb}{B} \quad (1\text{-}8(\text{b}))$$

$$l = \frac{nc}{C} \quad (1\text{-}8(\text{c}))$$

In these expressions, n is the factor that may be required to reduce h, k, and l to integers.

An intercept on the negative side of the origin is indicated by a bar or minus sign positioned over the appropriate index. Furthermore, reversing the directions of all indices specifies another plane parallel to, on the opposite side of, and equidistant from the origin. Several low-index planes are represented in Fig. 1-10.

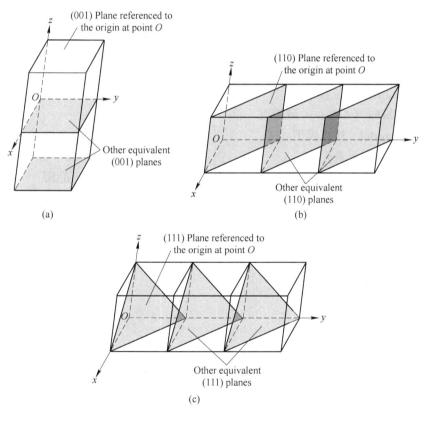

Fig. 1-10 Representations of a series each of the (a) (001), (b) (110), and (c) (111) crystallographic planes

One interesting and unique characteristic of cubic crystals is that planes and directions having the same indices are perpendicular to one another; however, for other crystal systems there are no simple geometrical relationships between planes and directions having the same indices.

Words and Expressions

microstructure 微观组织
metallic 金属
crystal 晶体
crystallographic 晶体学的
hard-sphere model 钢球模型
radii 半径（radius 的复数）
face-centered cubic 面心立方
body-centered cubic 体心立方
hexagonal close-packed 密排六方
unit cell 晶胞
aggregate 集合
diagonal 对角线
cube edge length 立方棱长（晶格常数）
coordination number 配位数
atomic packing factor（APF） 致密度
reduced-sphere model 质点模型
chromium 铬（化学元素）
tungsten 钨
polonium 钋
metalloid 类金属
symmetry 对称

crystal system 晶系
interaxial angle 轴间角
lattice 晶格
tetragonal 四方的
orthorhombic 正交的
rhombohedral 菱形的
monoclinic 单斜的
triclinic 三斜的
crystallographic direction 晶向
indices 指数（index 的复数）
vector 矢量
integer value 整数值
antiparallel 反平行的
positive and negative 正负
basal plane 基面
perpendicular 垂直的
orientations 取向
crystallographic plane 晶面
Miller indices 密勒指数
intercept 截距
reciprocal 倒数

2 Strengthening Mechanisms

2.1 Mechanisms of Strengthening in Metals

Metallurgical and materials engineers are often called on to design alloys having high *strengths* yet some *ductility and toughness*; typically, ductility is usually decreased when an alloy is strengthened. Several hardening techniques are frequently selected depending on the capacity of a material to be tailored with the *mechanical performance* required for a particular application.

Important to the understanding of strengthening mechanisms is the relation between *dislocation* motion and mechanical behavior of metals. Because macroscopic *plastic deformation* corresponds to the motion of large numbers of dislocations, the ability of a metal to deform plastically depends on the ability of dislocations to move. Because *hardness* and strength (both *yield and tensile*) are related to the ease with which plastic deformation can be made to occur, by reducing the mobility of dislocations, the mechanical strength may be enhanced, that is, greater mechanical forces are required to initiate plastic deformation. In contrast, the more unconstrained the dislocation motion, the greater is the facility with which a metal may deform, and the softer and weaker it becomes. Virtually all strengthening techniques rely on this simple principle: Restricting or hindering dislocation motion renders a material harder and stronger.

The present discussions focus on strengthening mechanisms for *single-phase metals* by *grain size* reduction, *solid-solution* alloying, and *strain hardening*. Deformation and strengthening of *multiphase alloys* are more complicated, involving concepts beyond the scope of the present discussion.

2.1.1 Strengthening by Grain Size Reduction

The size of the grains, or average grain diameter, in a *polycrystalline metal* influences the mechanical properties. Adjacent grains normally have different *crystallographic orientations* and, of course, a common *grain boundary*, as indicated in Fig. 2-1. During plastic deformation, *slip* or dislocation motion must take place across this common boundary—say, from grain A to grain B in Fig. 2-1. The grain boundary acts as a barrier to dislocation motion for two reasons:

(1) Because the two grains are of different orientations, a dislocation passing into grain B must change its direction of motion; this becomes more difficult as the crystallographic *misorientation* increases.

(2) The atomic disorder within a grain boundary region results in a discontinuity of slip planes from one grain into the other.

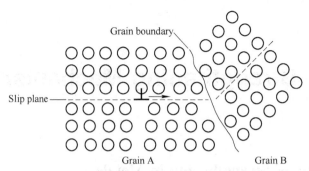

Fig. 2-1 The motion of a dislocation as it encounters a grain boundary, illustrating how the boundary acts as a barrier to continued slip. Slip planes are discontinuous and change directions across the boundary

It should be mentioned that, for *high-angle grain boundaries*, it may not be the case that dislocations traverse grain boundaries during deformation; rather, dislocations tend to "*pile up*" (or back up) at grain boundaries. These pile-ups introduce stress concentrations ahead of their slip planes, which generate new dislocations in adjacent grains.

A fine-grained material (one that has small grains) is harder and stronger than one that is coarse grained because the former has a greater total grain boundary area to impede dislocation motion. For many materials, the yield strength σ_y varies with grain size according to

$$\sigma_y = \sigma_0 + k_y d^{-1/2} \tag{2-1}$$

In this expression, termed the *Hall-Petch equation*, d is the average grain diameter, and σ_0 and k_y are constants for a particular material. Note that Eq. (2-1) is not valid for both very large (i.e., coarse) grain and extremely fine grain polycrystalline materials. Fig. 2-2 demonstrates the yield strength dependence on grain size for a brass alloy. Grain size may be controlled by the rate of *solidification* from the liquid phase, and also by plastic deformation followed by an appropriate *heat treatment*.

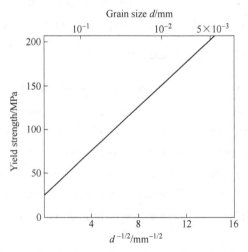

Fig. 2-2 The influence of grain size on the yield strength of a 70 Cu-30 Zn brass alloy. Note that the grain diameter increases from right to left and is not linear

It should also be mentioned that grain size reduction improves not only the strength, but also the toughness of many alloys.

Small-angle grain boundaries are not effective in interfering with the slip process because of the slight crystallographic difference across the boundary. However, *twin boundaries* effectively block slip and increase the strength of the material. Boundaries between two different phases are also impediments to movements of dislocations; this is important in the strengthening of more complex alloys. The sizes and shapes of the constituent phases significantly affect the mechanical properties of multiphase alloys.

2.1.2 Solid-solution Strengthening

Another technique to strengthen and harden metals is alloying with *impurity atoms* that go into either *substitutional* or *interstitial solid solution*. Accordingly, this is called *solid-solution strengthening*. High-purity metals are almost always softer and weaker than alloys composed of the same base metal. Increasing the concentration of the impurity results in an increase in tensile and yield strengths, as indicated in Fig. 2-3(a) and Fig. 2-3(b), respectively, for copper-nickel alloys; the dependence of ductility on nickel concentration is presented in Fig. 2-3(c).

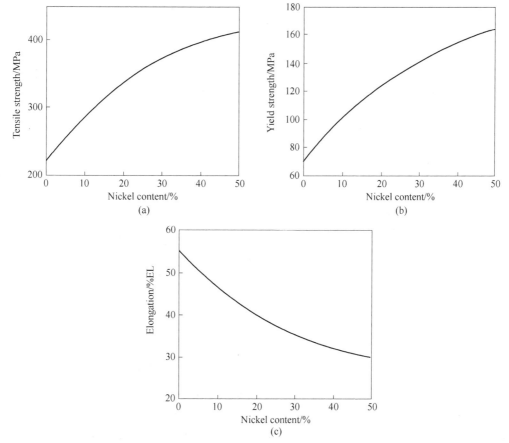

Fig. 2-3 Variation with nickel content of (a) tensile strength, (b) yield strength, and (c) ductility (%EL) for copper-nickel alloys, showing strengthening

Alloys are stronger than pure metals because impurity atoms that go into solid solution typically impose lattice strains on the surrounding host atoms. *Lattice strain* field interactions between dislocations and these impurity atoms, consequently, dislocation movement is restricted. For example, an impurity atom that is smaller than a host atom for which it replaces exerts tensile strains on the surrounding crystal lattice, as illustrated in Fig. 2-4(a). Conversely, a larger substitutional atom imposes compressive strains on its adjacent host atoms (Fig. 2-5(a)). These solute atoms tend to *diffuse* to and segregate around dislocations in such a way as to reduce the overall strain energy, that is, to cancel some of the strain in the lattice surrounding a dislocation. To accomplish this, a smaller impurity atom is located where its tensile strain partially counteracts some of the dislocation's compressive strain. For the *edge dislocation* in Fig. 2-4(b), this would be adjacent to the dislocation line and above the slip plane. A larger impurity atom would be situated as in Fig. 2-5(b).

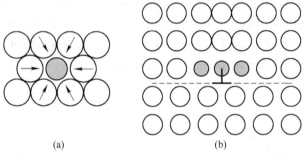

Fig. 2-4 (a) Representation of tensile lattice strains imposed on host atoms by a smaller substitutional impurity atom; (b) Possible locations of smaller impurity atoms relative to an edge dislocation such that there is partial cancellation of impurity-dislocation lattice strains

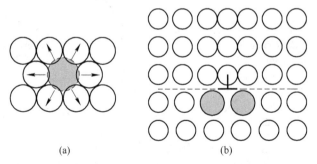

Fig. 2-5 (a) Representation of compressive strains imposed on host atoms by a larger substitutional impurity atom; (b) Possible locations of larger impurity atoms relative to an edge dislocation so that there is partial cancellation of impurity-dislocation lattice strains

The resistance to slip is greater when impurity atoms are present because the overall lattice strain must increase if a dislocation is torn away from them. Furthermore, the same lattice strain interactions (Fig. 2-4(b) and Fig. 2-5(b)) exist between impurity atoms and dislocations in motion during plastic deformation. Thus, a greater applied stress is necessary to first initiate and then continue plastic deformation for solid-solution alloys, as opposed to *pure metals*; this is evidenced by the enhancement of strength and hardness.

2.1.3 Strain Hardening

Strain hardening is the phenomenon by which a ductile metal becomes harder and stronger as it is plastically deformed. Sometimes it is also called *work hardening*, or, because the temperature at which deformation takes place is "cold" relative to the absolute melting temperature of the metal, cold working. It is sometimes convenient to express the degree of plastic deformation as percent cold work rather than as strain. Percent cold work (%CW) is defined as

$$\%\mathrm{CW} = \left(\frac{A_0 - A_\mathrm{d}}{A_0}\right) \times 100 \qquad (2-2)$$

where A_0 is the original area of the cross section that experiences deformation; A_d is the area after deformation.

Fig. 2-6(a) and Fig. 2-6(b) demonstrate how steel, brass, and copper increase in yield and tensile strength with increasing cold work. The price for this enhancement of hardness and strength is in the ductility of the metal. This is shown in Fig. 2-6(c), in which the ductility, in percent elongation, experiences a reduction with increasing percent cold work for the same three alloys. The influence of cold work on the stress-strain behavior of *a low-carbon steel* is shown in Fig. 2-7; here, *stress-strain curves* are plotted at 0%CW, 4%CW, and 24%CW.

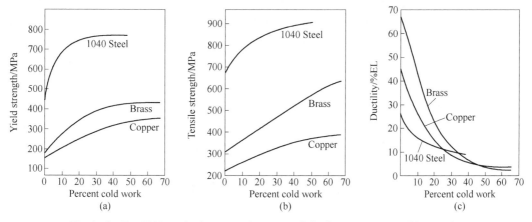

Fig. 2-6 For 1040 steel, brass, and copper, (a) the increase in yield strength, (b) the increase in tensile strength, and (c) the decrease in ductility with percent cold work

The strain-hardening phenomenon is explained on the basis of dislocation-dislocation strain field interactions. The dislocation density in a metal increases with deformation or cold work because of *dislocation multiplication* or the formation of new dislocations. Consequently, the average distance of separation between dislocations decreases and the dislocations are positioned closer together. Generally, dislocation-dislocation strain interactions are repulsive. The net result is that the motion of a dislocation is hindered by the presence of other dislocations. As the dislocation density increases, this resistance to dislocation motion by other dislocations becomes more

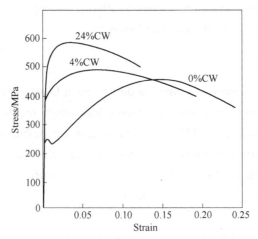

Fig. 2-7 The influence of cold work on the stress-strain behavior of a low-carbon steel; curves are shown for 0%CW, 4%CW, and 24%CW

pronounced. Thus, the imposed stress necessary to deform a metal increases with increasing cold work.

Strain hardening is often utilized commercially to enhance the mechanical properties of metals during fabrication procedures. The effects of strain hardening may be removed by an *annealing* heat treatment. *Strain-hardening exponent* which is a measure of the ability of a metal to strain harden; the larger its magnitude, the greater is the strain hardening for a given amount of *plastic strain*.

In summary, we have discussed the three mechanisms that may be used to strengthen and harden single-phase metal alloys: strengthening by grain size reduction, solid-solution strengthening, and strain hardening. Of course, they may be used in conjunction with one another; for example, a solid-solution strengthened alloy may also be strain hardened.

It should also be noted that the strengthening effects due to grain size reduction and strain hardening can be eliminated or at least reduced by an elevated-temperature heat treatment. In contrast, solid-solution strengthening is unaffected by heat treatment.

2.2 Recovery, Recrystallization, and Grain Growth

Plastically deforming a polycrystalline metal specimen at temperatures that are low relative to its absolute melting temperature produces microstructural and property changes that include a change in *grain shape*, strain hardening, and an increase in dislocation density. Some fraction of the energy expended in deformation is stored in the metal as *strain energy*, which is associated with tensile, *compressive*, and *shear* zones around the newly created dislocations. Furthermore, other properties, such as *electrical conductivity* and *corrosion resistance*, may be modified as a consequence of plastic deformation.

These properties and structures may revert back to the precold-worked states by appropriate heat treatment (sometimes termed an annealing treatment). Such restoration results from two different processes that occur at elevated temperatures: *recovery and recrystallization*, which may be followed by *grain growth*.

2.2.1 Recovery

During recovery, some of the stored internal strain energy is relieved due to dislocation motion (in the absence of an externally applied stress), as a result of enhanced atomic diffusion at the elevated temperature. There is some reduction in the number of dislocations, and dislocation configurations are produced having low strain energies. In addition, *physical properties* such as electrical and *thermal conductivities* recover to their precold-worked states.

2.2.2 Recrystallization

Even after recovery is complete, the grains are still in a relatively high strain energy state. Recrystallization is the formation of a new set of strain-free and *equiaxed grains* (i. e., having approximately equal dimensions in all directions) that have low dislocation densities and are characteristic of the precold-worked condition. The *driving force* to produce this new grain structure is the difference in internal energy between the strained and unstrained material. The new grains form as very small *nuclei* and grow until they completely consume the parent material, processes that involve *short-range diffusion*. Several stages in the recrystallization process are represented in Fig. 2-8(a) to Fig. 2-8(d); in these photomicrographs, the small speckled grains are those that have recrystallized. Thus, recrystallization of cold-worked metals may be used to refine the grain structure.

Also, during recrystallization, the mechanical properties that were changed as a result of *cold working* are restored to their precold-worked values, that is, the metal becomes softer and weaker, yet more ductile. Some heat treatments are designed to allow recrystallization to occur with these modifications in the mechanical characteristics. The extent of recrystallization depends on both time and temperature. The degree (or fraction) of recrystallization increases with time, as may be noted in the photomicrographs shown in Fig. 2-8(a) to Fig. 2-8(d).

(a)

(b)

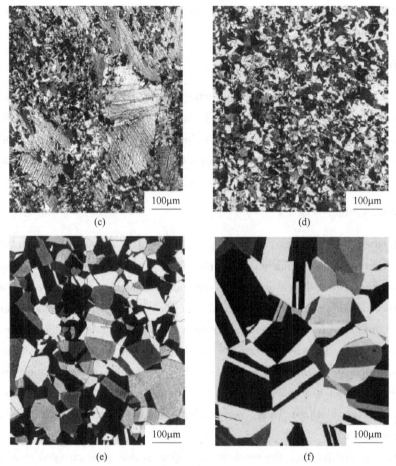

Fig. 2-8 Photomicrographs showing several stages of the recrystallization and grain growth of brass
(a) Cold-worked (33%CW) grain structure; (b) Initial stage of recrystallization
after heating for 3s at 580℃; the very small grains are those that have recrystallized;
(c) Partial replacement of cold-worked grains by recrystallized ones (4s at 580℃);
(d) Complete recrystallization (8s at 580℃); (e) Grain growth after 15min at 580℃;
(f) Grain growth after 10min at 700℃

The influence of temperature is demonstrated in Fig. 2-9, which plots tensile strength and ductility (at room temperature) of a brass alloy as a function of the temperature and for a constant heat treatment time of 1h. The grain structures found at the various stages of the process are also presented schematically.

The recrystallization behavior of a particular metal alloy is sometimes specified in terms of a recrystallization temperature, the temperature at which recrystallization just reaches completion in 1 h. Thus, the recrystallization temperature for the brass alloy of Fig. 2-9 is about 450℃. Typically, it is between one-third and one-half of the absolute *melting temperature* of a metal or alloy and depends on several factors, including the amount of prior cold work and the purity of the alloy. Increasing the percent cold work enhances the rate of recrystallization, with the result that the recrystallization temperature is lowered, and approaches a constant or limiting value at high deformations; this effect is shown in Fig. 2-10. Furthermore, it is this limiting or minimum recrystallization temperature that is normally specified in the literature. There exists some *critical degree of cold work* below which recrystallization cannot be made to occur, as shown in the figure; typically, this is between 2% and 20% cold work.

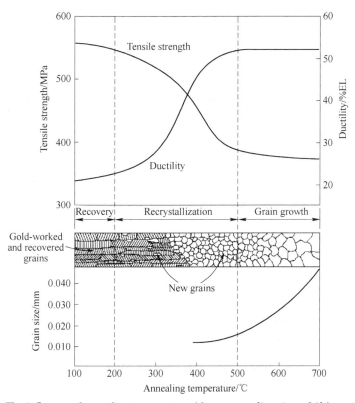

Fig. 2-9 The influence of annealing temperature (for an annealing time of 1h) on the tensile strength and ductility of a brass alloy. Grain size as a function of annealing temperature is indicated. Grain structures during recovery, recrystallization, and grain growth stages are shown schematically

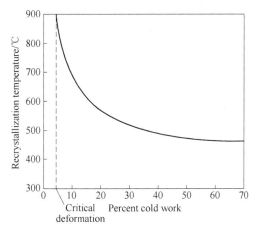

Fig. 2-10 The variation of recrystallization temperature with percent cold work for iron, for deformations less than the critical (about 5%CW), recrystallization will not occur

Recrystallization proceeds more rapidly in pure metals than in alloys. During recrystallization, grain-boundary motion occurs as the new grain nuclei form and then grow. It is believed that impurity atoms preferentially segregate at and interact with these recrystallized grain boundaries so as to diminish their (i.e., grain boundary) mobilities; this results in a decrease of the recrystallization rate and raises the recrystallization temperature, sometimes quite substantially. For pure metals, the recrystallization temperature is normally $0.4T_m$, where T_m is the absolute melting temperature; for some commercial alloys it may run as high as $0.7T_m$. Recrystallization and melting temperatures for a number of metals and alloys are listed in Table 2-1.

Table 2-1 Recrystallization and melting temperatures for various metals and alloy

Metal	Recrystallization Temperature/℃	Melting Temperature/℃
Lead	-4	327
Tin	-4	232
Zinc	10	420
Aluminum (99.999%)	80	660
Copper (99.999%)	120	1085
Brass (60 Cu-40 Zn)	475	900
Nickel (99.99%)	370	1455
Iron	450	1538
Tungsten	1200	3410

It should be noted that because recrystallization rate depends on several variables, as discussed previously, there is some inadequacy to recrystallization temperatures cited in the literature. Furthermore, some degree of recrystallization may occur for an alloy that is heat treated at temperatures below its recrystallization temperature.

Plastic deformation operations are often carried out at temperatures above the recrystallization temperature in a process termed hot working. The material remains relatively soft and ductile during deformation because it does not strain harden, and thus large deformations are possible.

2.2.3 Grain Growth

After recrystallization is complete, the strain-free grains will continue to grow if the metal specimen is left at the elevated temperature (Fig. 2-8(d) to Fig. 2-8(f)); this phenomenon is called grain growth. Grain growth does not need to be preceded by recovery and recrystallization; it may occur in all polycrystalline materials, metals and *ceramics* alike.

An energy is associated with grain boundaries. As grains increase in size, the total boundary area decreases, yielding an attendant reduction in the total energy; this is the driving force for grain growth.

Grain growth occurs by the *migration of grain boundaries*. Obviously, not all grains can enlarge, but large ones grow at the expense of small ones that shrink. Thus, the average grain size increases with time, and at any particular instant there exists a range of grain sizes. Boundary motion is just the short-range diffusion of atoms from one side of the boundary to the other. The directions of boundary movement and atomic motion are opposite to each other, as shown in Fig. 2-11.

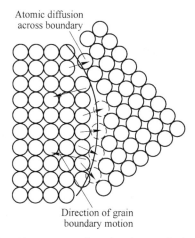

Fig. 2-11 Schematic representation of grain growth via atomic diffusion

For many polycrystalline materials, the grain diameter d varies with time t according to the relationship

$$d^n - d_0^n = Kt \qquad (2-3)$$

where d_0 is the initial grain diameter at $t=0$; K and n are time-independent constants; the value of n is generally equal to or greater than 2.

The dependence of grain size on time and temperature is demonstrated in Fig. 2-12, a plot of the logarithm of grain size as a function of the logarithm of time for a brass alloy at several temperatures. At lower temperatures the curves are linear. Furthermore, grain growth proceeds more rapidly as temperature increases—that is, the curves are displaced upward to larger grain sizes. This is explained by the enhancement of diffusion rate with rising temperature.

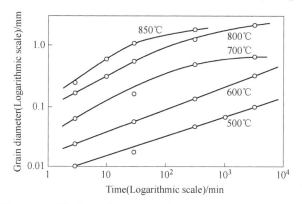

Fig. 2-12 The logarithm of grain diameter versus the logarithm of time for grain growth in brass at several temperatures

The mechanical properties at room temperature of a fine-grained metal are usually superior (i.e., higher strength and toughness) to those of coarse-grained ones. If the grain structure of a single-phase alloy is coarser than that desired, *refinement* may be accomplished by plastically deforming the material, then subjecting it to a recrystallization heat treatment, as described previously.

Words and Expressions

metallurgical 冶金学的	diffuse 扩散
strength 强度	edge dislocation 刃型位错
ductility 伸长率	pure metals 纯金属
toughness 韧性	work hardening 加工硬化
mechanical performance 机械性能	a low-carbon steel 低碳钢
dislocation 位错	stress-strain curves 应力应变曲线
plastic deformation 塑性变形	dislocation multiplication 位错增殖
hardness 硬度	annealing 退火
yield and tensile 屈服和拉伸强度	strain-hardening exponent 应变硬化指数
single-phase metals 单相金属	plastic strain 塑性应变
grain size 晶粒尺寸	grain shape 晶粒形貌
solid-solution 固溶	strain energy 应变能
strain hardening 应变硬化	compressive 压缩
multiphase alloys 多相合金	shear 剪切
polycrystalline metal 多晶金属	electrical conductivity 导电性
crystallographic orientation 晶体取向	corrosion resistance 耐腐蚀性
grain boundary 晶界	recovery 回复
slip 滑移	recrystallization 再结晶
misorientation 位相差	grain growth 晶粒长大
high-angle grain boundary 大角度晶界	physical properties 物理性能
pile up 堆积	thermal conductivity 导热性
Hall-Petch equation 霍尔-佩奇方程	equiaxed grains 等轴晶
solidification 凝固	driving force 驱动力
heat treatment 热处理	nuclei 胚核
small-angle grain boundary 小角度晶界	short-range diffusion 短程扩散
twin boundary 孪晶界	cold working 冷加工
impurity atoms 杂质原子	melting temperature 熔点温度
substitutional solid solution 置换固溶	critical degree of cold work 临界变形量
interstitial solid solution 间隙固溶	ceramics 陶瓷
solid-solution strengthening 固溶强化	migration of grain boundary 晶界迁移
lattice strain 晶格应变	refinement 晶粒细化

3 The History of Rolling

3.1 The Early History of Rolling

In its earliest beginnings, the *rolling* of flat materials was undoubtedly limited to those metals of sufficient *ductility* to be worked *cold*, and it is probable that it was first performed by *goldsmiths* or those manufacturing *jewelry* or *works of art*. Yet, as it is the case with many other important processes, metal rolling cannot be traced to a single inventor.

During the 14th century, small hand-driven *rolls* about 13mm in diameter were used to flatten *gold* and *silver* and perhaps *lead*. However, the first true *rolling mills* of which any record exists were designed by Leonardo da Vinci in 1480. Sketches in his notebook show two mills, driven by *worm gears*, for rolling *lead sheets* and also a machine for producing *tapered lead bars* by means of a die and a roll. Yet, there is no evidence that these mills were ever built, and there is a fair degree of certainty that metal rolling was not of any importance before the middle of the 16th century.

Before the end of the 16th century, however, at least two mills *embodying* the basic ideas of rolling are known to have been in operation. A Frenchman named Brulier in 1553 rolled sheets of gold and silver to obtain uniform *thickness* for making coins and mills for rolling *mint flats* were in use in 1581 at the Pope's *mint*, in 1587 in Spain, and in 1599 in Florence. In 1578, Bevis Bulmer received a *patent* for the operation of a *slitting mill* which consisted of a series of *discs* mounted on two *spindles*, one above the other, in such a manner that the *flat bar* passing between the *revolving* discs was cut into *strip*. A mill of this type was set up at Dartford, in Kent, in 1590, by Godefroi de Bochs, a native of Liege, Belgium.

During the same period, lead was also beginning to find increasing use for *roofing*, for *flashing*, and for other purposes. Salomon de Caus of France, in 1613 built a hand-operated mill for rolling sheets of lead and *tin* used in making organ *pipes*, the rolls being turned by a "*strong-armed cross*" attached to the lower *axle*.

With the exception of the Bulmer slitting mill, mentioned above, all of these early developments *pertain to* the rolling of *softer metals*, *presumably* at ambient temperatures. Johannsen, in "Geschichte des Eisens" says, "The use of rolls in an iron works was a German development of the 16th century. Belgium and England both started to use rolls about the same time, and they are both sometimes cited as the *birthplace* of rolling." All three nations probably reached this development at about the same time, but there is little evidence of anything other than slitting mills in the 16th century, and still less evidence to give any nation a clear *claim* to *priority*. Such information as is available indicates that, in the rolling of iron, Great Britain *led the way*. No record of development during the first half of the 17th century exists, but we know that in 1665 a rolling mill was in operation in the Parish of Bitton, near Bristol, and it is stated that, from 1666 on, iron was rolled into *thin flats* for *slitting*.

The rolling of bars was *foreshadowed* during this period but was not brought to *fruition*. In 1679 a patent was issued covering the *finishing* of bolts by rolling, and in 1680 bars were being passed through *plane-surfaced rolls* to *flatten out irregularities*. A *pamphlet* on the "British Iron Trade" published in 1725 states, however, that even at that date all bars were *hammered*.

However, by 1682, *large rolling mills* for the *hot rolling* of *ferrous materials* were in operation at Swalwell and Winlaton, near Newcastle, England. Using these mills, bars were rolled into sheets and the sheets cut into *rods* at the slitting mills. Soon after this date, at Pontypool in Wales, John Hanbury began using at his *ironworks* a rolling mill as an independence machine for the production of thin sheet iron. Edward Llwyd, in a letter dated June 15, 1697 wrote, "Hanbury showed us an excellent invention of his own, for driving hot iron (by the help of a rolling machine moved by mater) into as thin *plates*. They cut their common iron bars into pieces of about 610mm long, and heating them glowing hot, place them between these iron rollers. The rollers, moved with water, drive out these bars to such thin plates, which was about 100mm long."

Although Major Hanbury designed the rolling mill described in the letter, there is no evidence that he originated the idea of hot-rolling bars into thin sheets since it is believed that the practice was general throughout Europe by 1660 being known in German very early in the century. At any rate, Germany *monopolized* the growing English market for *tinplate* from shortly after 1620 until Major Hanbury began tinplate manufacture in this same Pontypool works sometime before 1720. After that, for more than 150 years, Wales was the major source of tinplate and *terne plate*.

During the early part of the 18th century, there is no doubt that rolling mills were in common used both in England and on the continent. Christopher Polhem (1720~1746), Sweden's great mechanical genius, wrote of rolling mills at about this time and his writing indicates that he assumed his readers were familiar with them. Polhem himself designed a mill very similar to the modern *Lauth mill*, except that his mill utilized four rolls, with the *backup rolls* driven.

A machine for rolling sheets of lead, that really foretold the shape of subsequent rolling mills, was brought to France from England in 1728. This mill used rolls 1520mm long and 300mm in diameter and was *equipped with* a *roller table* 7320mm long at front and back. It was a radical departure from the accepted mill design *of the day*, being a *reversing mill*, controlled by a *clutch* and *gear system*. The *plain rolls* could be replaced with other, *grooved rolls* 400mm in diameter, with grooves ranging from 100mm to 50mm in diameter, with which hollow *cast lead ingots* were rolled into pipe over a *mandrel*.

In 1728, a patent for a mill to roll hammered bars "into such shapes and forms as shall be required" was issued to John Payne in England. However, Payne's concepts do not appears to have been reduced to practice, but the rolling of iron bars and shapes was of interest to *steelmakers* and appears to have been practiced. For example, in 1747, the Academic des Cinces appointed a commission to visit a new mill at Essonne, France, which rolled iron bars. By this time, the practice of rolling iron plates for tinning by *cold rolling* was also *in vogue*. In 1759, a patent was *granted* to Thomas Blockley of England for "rolling metals" —a broad description about rolls which the user could groove to suit his requirements; and in 1766, another Englishman, John Purnell received a patent for grooved rolls with *coupling boxes* and *nut pinions* for turning the rolls *in unison*. Until this time, rolls were individually driven, and the unequal *rates of revolution* caused excessive *wear*, as well as making it necessary to install *guides* on both sides of each roll.

In the manufacture of *hand-forged plates* for tinplating, it had been the practice to hammer several layers of metal at one time. Accordingly, when the rolling mill came into use for producing sheets for tinning, it became the practice, about 1756, to double the sheet after some *elongation* and then proceed with the further rolling of the two thickness of metal. Sometimes in this "*pack rolling*" or "*ply rolling*" the doubling was repeated so that four thickness of steel were rolled simultaneously.

The aforementioned hot-rolled mills consisted essentially of a *single-stand* of *two rolls*, one above the other. As described by Shannon, the operating principle of this type of *hot mill* consists briefly of the following steps: heat the *piece*, pass the piece through the rolls, push the piece back over the *top roll* with *hand tongs*, pass the piece through the rolls again, and so on until the piece being rolled either is of the required thickness or else has *cooled down* to the point where it must be *reheated* before the rolling can continue. This is only a *bare outline* of the elements of the operation, which is very much further complicated by adjusting of the rolls, and various other factors, according to the nature of the sheets being rolled. *Owing to* the fact that the piece is fed to the rolls by the roller and is caught on the other side and handed back to the roller for passing through the rolls again, and since there are only two rolls, one above the other in set, the *conventional sheet hot mill* or *sheet mill* is described as a "*two-high, pull-over mill*". More than one of these mills may be employed in the *rolling unit*, each performing a separate stage of the work, but the principle of each remains the same."

The 18th century also saw the advent of the *tandem mill* in which the metal rolled in *successive stands*. The first true tandem mill of which we have record was patented in England by Richard Ford, in 1766, for the hot-rolling of *wire rods*. James Cockshutt and Richard Crawshay, about 1790, erected a *four-high tandem mill* near Sheffield, England. This mill was about 1520mm in length and less than 610mm high, with a capacity of, probably, one or two tons per day. A later patent, issued in 1798, refers to a tandem mill for rolling of plates and sheets. In the same year, John Hazeldine added *mechanical guides* to a *rod mill*.

3.2　Late Developments in the Hot Rolling of Steel in Europe

The advent of tandem rolling practice may, however, be said to date from 1783 when a patent was granted to Henry Cort of Fontley Iron Mills, near Fareham, England, for utilizing grooved rolls for rolling iron bars. A mill with rolls of this design could produce at least 15 times the output per day obtaining with a *tilt hammer*. However, the claim to invention put forward by Cort and his *successors* was strenuously *contested* at a later date. He was not the first to use grooved rolls, but he was the first to combine the use of all the best features of the various steelmakings known at that time. This fact alone justifies the term "father of modern rolling", which has been applied to him by modern writers.

In the beginning of the 19th century, the *industrial revolution* in England was gathering *momentum*, creating an unprecedented demand for *iron and steel*. Accordingly, rolling mill developments were numerous and important. John Birkenshaw started the first *rail rolling mill* in 1820 producing *wrought iron* rail in lengths of 4570mm to 5490mm. In 1831 the first *I-beam* were rolled by Zores in Paris in 1849.

Both the sizes of the mills and the sizes of rolled product grew rapidly. At *the British Great Exposition* of 1851, a *plate* 6100mm long, 1070mm wide and 11mm thick was exhibited by the Consett Iron Company. This plate weighted 510 kg and was the largest plate rolled up to that time.

Three-high mills were also introduced about the middle of the century. A British patent for such a mill designed for rolling *heavy sections* was granted in 1853 to R. B. Roden of the Abersychen Iron works. In this mill, the middle roll was driven and fixed in the *housing* while the *upper* and *lower rolls* were adjustable in position. On the same mill, a steam-operated *lifting table* raised and lowered the materials to be rolled. This design was improved on a few years later by Bernard Lauth who used a *middle roll* of smaller diameter than the upper and lower rolls, as illustrated in Fig. 3-1. This modification to the mill provided it with a higher productivity with less power utilization.

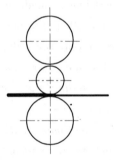

Fig. 3-1 Lauth's mill with smaller middle roll

In mid century, the first *reversing plate mill* was put into operation at the Parkgate Works in England, and in 1854 it was used to roll the plates for the "Great Eastern" *steamship*. In 1848, the *universal mill* was invented by Daelen R M of Lenderdorf, Germany, who built the first mill of this type about seven years later. Although patents for "*continuous*" *hot mills* were issued to Sir Henry Bessemer in 1857 and to Dr. Leach R V in 1859, "the first mill constructed on the continuous principle of rolling iron or steel" was the subject of a patent issued to Charles While of Pontypridd, Wales.

However, an apparently more successful continuous mill was patented in 1862 by George Bedson of the Bradford Iron Works at Manchester, England, in which he claims the employment of a series of rolls placed at varying angles, whereby the necessity of turning the metal is avoided. This was a *rod rolling mill* in which a 45 kg *billet* of 685mm² cross section was drawn through 16 pairs of rolls *in line*, 8 *horizontal* and 8 *vertical*. Its production rate was such that 20 tons of wire rods could be rolled in 10 hours.

A British patent issued in 1862 to J. T. Newton of Ystalyfeva, Wales, described *predecessor* to the modern *cluster mill*, in as much as it used small *work rolls* backed up by other of larger diameter. The work rolls were driven but the pressure was applied by the larger backup rolls, a principle utilized in both the hot and cold mills of today.

The four-high mill with its rolls in the same vertical plane was introduced by Bleckley of Warrington, England, in 1872, to finish wrought-iron workpieces from which rails were rolled. Mills to produce *Z-bars* were put in use in Germany in 1863, and in 1867, *beams* 200mm to 300mm deep were rolled on a mill designed by Menelaus, of the Dowlais Works in Wales. This mill contained two pairs of rolls, one pair placed in a vertical plane, somewhat higher than the other pair of rolls. Rive-de-Giev, France, was rolling beams on a universal mill in 1872, and four years later, Joseph de Buigne, of France, rolled the first *H-beams* produced on a continuous mill.

Tandem rolling of hot steel took an *upsurge* around 1890, and in 1892, a *semi-continuous hot strip mill*, with a mechanically geared *two-high tandem finishing train*, was built at Teplitz, Bohemia. It was reported to have rolled sheets up to 1270mm in *width*, in *thickness* from 2mm to 3mm and in *length* up to 18290mm. The mill utilized a *roughing train* of *two three-high stands* and a *finishing train* consisting of *five-stand* of 620mm by 1500mm rolls. Each train was powered by a 750kW engine. Since the works at Teplitz were abandoned in 1907, it is to be assumed that the mill was not a *commercial* success.

3.3 The Early Rolling of Steel in the United States

To all intents and purposes, the history of metalworking in the United States began with the arrival of colonists from Europe. Since skilled *metalworkers* were present in every colony, the colonists supplied a great part of their own needs for metals. However, in 1750, *the British Parliament* decreed that "no mill or other engine for slitting or rolling iron, or any furnace for making steel" should be built in the American Colonies. This law was generally *disregarded* so that by 1775, the colonies were producing 30000 tons of iron per year, only one third of which was exported to England as *pig iron*.

The first American rolling mill was built in 1751 for Peter Oliver in the province of Middlesboro, Massachusetts. It was used to roll down 76mm wide *hammered bars* made at the *forges* from a thickness of about 19mm to 6mm suitable for slitting into *nail rods* in 4 *passes*. The rolls were each driven by a *waterwheel* 5490mm in diameter. The *chilled iron rolls* were 910mm long by 380mm in diameter and were designed with *roll necks* 230mm in diameter.

At the outbreak of *the American Revolution*, the colonies possessed a flourishing iron industry from which all restrictions were ended with the establishment of independence. Unhampered by tradition, the Americans succeeded, however, within the next two centuries to develop the largest national steel industry in the world.

Some of the more notable steps in the development of the new steel industry were as follows. Issac Pennoch established a slitting mill on Bush's Run in 1793, and by 1810 this plant was rolling plates with mills using rolls 410mm to 460mm in diameter and 910mm to 1220mm long, driven by an waterwheel. In 1820, Dr. Charles Lukens, Pennoch's son-in-law, rolled *boiler plate* for the first time at this plant which eventually developed into the present Lukens *Steel Corporation*.

However, at the beginning of the 19th century, it was apparent that the Pittsburgh area was becoming the focal point of the industry. Christopher Crown built the first rolling mill in Western Pennsylvania, and, *incidentally*, the first one known to have been powered by steam. It utilized a 50kW steam engine which also supplied power for a slitting mill and a tilt hammer. At Plumsock, Issac Meason, in 1816, built the first American mill for rolling flat bars. Two years later, the Pittsburgh Steam Engine Company built a sheet rolling mill, and in 1819, the first *angle iron* rolled in the United States was produced at the Union Rolling Mill in Pittsburgh. By 1825 five rolling mills were in operation in Pittsburgh and a sixth was under construction.

By the middle of the 19th century, iron production in the USA had risen to 350000 tons per year and the increasing availability of metal *fostered* considerable *inventiveness* designed to further process and fabricated metal parts. The rolling of *corrugated plates* was patented in 1850, and in the same year an Ohioan patented an improved machine for rolling irregular forms of metal such as spades, shovels, axes, and hammers.

The first three-high beam mill in the USA was used by the Trenton Iron Works, at Trenton, built in 1852, it employed three vertical rolls. Another beam mill departing radically from previous designs was supposed to have been built in 1853 by Charles Hewitt of Trenton. However, the first thoroughly satisfactory three-high mill is generally attributed to John Fritz, who built such a mill for the Cambria Iron Works, Johnstown, Pa. in 1857 to roll tails between 460mm diameter rolls. This mill is of interest because it established the practice of mounting the *mill housing* on heavy, *cast guide rails*. However, Jones and Lauth of Pittsburgh, Pa. introduced the Lauth mill to the U. S. in 1859 buying all *U. S. rights* from the British patent owner.

After *the Civil War* ended in 1865, the rapid expansion of the railroads proved to be a tremendous *stimulus* to the American *iron and steel industry*. The first steel rails were rolled in 1865 by the North Chicago Rolling Mill Company. In 1867, George Fritz (brother to John Fritz) started the first successful *blooming mill* operation in the U. S. and, in the same year, the first beams were produced in Pittsburgh on a 510mm structure mill. Also in the same year, Andrew Kloman, with the help of John Zimmer, built the first universal mill in Pittsburgh capable of rolling plates from 178mm to 610mm wide and from 7mm to 50mm thick. In 1877, Mackintosh Hemphill designed and installed a 762 cm *reversing blooming mill* for Schoenberger and Company (later the Schoenberger Works of the American Steel and Wire Company). This was the first mill of its kind in the Pittsburgh area and possibly the first in the United States. In 1881, the firm of Mackintosh-Hemphill, Inc. built the first rolling mill of wholly American construction, this being a *two-high reversing blooming mill* at the Pittsburgh Bessemer Works (the forerunner of the present Homestead Works of the U. S. Steel Corporation).

The rolling of sheet steel in the United States began around 1880, and it is recorded that a three-high mill with 2130mm long rolls at the Brandywine Rolling Mill was used for this purpose. Sometime later, the same company installed a *three-high roughing mill* and a *three-high finishing mill*, both using chilled iron rolls 864mm in diameter and 3050mm long.

3.4 Energy Source for Rolling Mills

The earliest mills were operated by handpower, usually by turning a strong-armed cross attached to either or both of the rolls. With such limited power available, the only materials that could be rolled were the softer metals such as gold, silver, tin, and lead.

Waterwheels were next used to turn mill rolls, and this was a convenient development since such wheels were widely used in iron works. The first mill likely to have been powered by a waterwheel was the slitting mill built at Dartford in Kent, England, in 1590 by Godefroi de Bochs under a patent granted in 1588 to Bevis Bulmer. In the four-high mill erected near Sheffield, England, about 1790, by James Cockshutt and Richard Crawshay, the top and bottom rolls were each driven by separate waterwheels.

The first commercially successful steam engine was invented in 1698 by Thomas Savery of England, who, incidentally, was the first to evaluate an engine in term of *horsepower*. However, his steam engine was not first used to drive the mill rolls but to pump back into the *reservoir* the water that had already passed over the waterwheel. However, a Boulton and Watt steam engine was used to power a rolling mill and slitting mill at John Wilkinson's Bradley Works and a stem engine was used to power a tinplate rolling mill in 1798.

Improvements to steam engines occurred rapidly at the beginning of the 19th century and they were soon commonly used for driving mills, the power being transmitted to the rolls by direct mechanical connections through *shafts*, *couplings*, and gears. In the latter half of the 19th century, there was a constant demand for larger and larger engines so that by 1875, engines were being built capable of delivering in excess of 750kW, some being as large as 220kW to 300kW units.

Initially *flywheels* found extensive use in powering of mills, particularly after the development of 3-high mills (such as Lauth mill) which required no *reversal*. It became apparent, however, towards the end of the 19th century, that two-high mills with drives capable of rapid reversal were to be preferred, especially in view of the fact that they could be quickly stopped if necessary.

Yet steam power, at its best, was costly and inefficient and it was fortunate that the latter part of the 19th century saw the development of *electric generators* and motors. Electric power could be conveniently transmitted over wire to motors directly attached to the mills. Some of the generators were driven by *internal combustion engines* as, for example, at the Gary Steel Works. This plant was designed in 1908 to be the first sizable steel mill built for the use of electric power and it had 15 *gas-driven* generators each capable of outputs up to 2,000 kilowatts. The size was increased a few years alter to 3,000 kilowatts equaling the largest *reciprocating steam-driven* generator at that time.

Even earlier, *direct current motors* had been installed to operate some smaller mills. In 1903, two 1120kW motors powered a high rail mill at the Edgar Thomson Works at Braddock, Pa., and the first reversing d-c drive motor was installed the same year on a 9140mm universal plate mill at South Works in Chicago.

Both gas and steam engines operated at relatively slow speeds. Other improvements in the generation and distribution of electrical power led to a steady conversion to electric motors in steel mills not only in the USA but throughout the world. The use of variable speed d-c motors on *main drive* began in the early 1940's and has been gaining popularity ever since. At the same time, the power utilized by mills has gradually increased so that some of the more recent hot mill stands are driven by *multi-armature motors* providing 8950kW. In the case of modern *cold reduction facilities*, the stands of wide sheet mills are typically powered by motors of 4480kW.

3.5 The Historical Development of Cold Rolling

In spite of the fact that the first rolling of metals was, in effect, cold rolling, the flat, cold *reduction* of iron and steel does not seem to have been successfully carried out until the end of the 18th century. However, it should be noted that cold rolling was practiced on tin plate in England as early as 1747 and, in 1783, in the same country, John Westwood proposed the cold reduction of *steel bands* for watch springs. From 1825 to 1860, due mainly to improvement in the manufacture of rolls, considerable amount of *high-carbon* flats were produce by cold rolling.

When the first cold-rolling operation was undertaken in the USA is uncertain. It appears, however, that the flattening of wire carried out by the Washburn and Moen Company, in Worcester, Massachusetts contributed the first commercial operation of this type.

The development of the cold-rolling of steel as a production process, however, gained real *impetus* only after the evolution of the *Lauth 3-high cold mill* with its smaller diameter middle roll. Yet the advantages of smaller diameter work rolls had been recognized much earlier for Christopher Polhem described a 4-high mill for flat, hot rolling using "slender" *wrought-iron work rolls*, supported by large *cast-iron backing rolls*. The commercialization of the Lauth cold mill, however, was carried out principally in America by the old American Iron and Steel Company of Pittsburgh.

As the superior properties of *cold rolled strip* became more and more appreciated, cold rolling spread even more widely both in this country and abroad, being practiced primarily on 2-*high mills*, although 4-roll and 6-roll cluster mills were later used in this country. The first 4-high mill for the cold rolling of steel was first used on an experimental basis as recently as 1927 by the Allegheny Ludlum Steel Corporation.

Improvements to *roll neck bearings* also contributed to the increasing use of cold reduction mills. Roller bearings were first used on 2-high cold mills as early as 1890, on the backup rolls of cluster-type cold mill in 1909 and on 4-high mills in 1926.

Reversing cold mills of the 2-high type were first used in Germany in the 1920's and of 4-high type in 1923. The first such cold mill in this country was installed at Gary in 1933.

The first record of *tandem cold rolling* of steel strip goes back to about 1904, when the West Leechburg Steel Company installed and operated a *2-high 4-stand tandem mill*, each stand being driven by a separate, adjustable speed d-c motor. Real tandem mill operation, with *tension between stands*, and a *tension reel*, was developed around 1915, on mills installed by Superior Steel Company and the Morris and Bailey Steel Company of Pittsburgh. And in 1926, the first 4-high, 4-stand tandem cold mill was put into operation by the American Rolling Mill Company at their Butler plant.

In the operation of both reversing and tandem mills, as the strip being handled got longer, and was rolled at higher and higher speeds, the matter of handling the material necessarily demanded attention. This occurred first in the hot rolling of wire rods and the first reels of record were used around 1860, being manually rotated by a boy, who visually *synchronized* the reel with the mill. The *cold strip reel* seems to have preceded the *hot strip reel* or coiler by some ten years, the first cold reels being built in Germany by August Schmitz Company, around 1893. The reels were well designed units. The first *high-tension* reel was patented in 1905 by Conklin W F, of Pittsburgh. Around 1920, the separate, electrically-driven reel, maintaining *constant tension* by *current control*, was developed jointly by the Superior Steel Company and the Westinghouse Electric Corporation.

The *concurrent* development of reels and the electric drive resulted in the development of *strip tension control*, first between mill and reel, and later between adjacent stands of the tandem mill. Although the first tandem cold mill was operated with *slack strip* between stands, the art was later advanced to the using of slack or *loop take-ups*, and finally to the adoption of high *interstand tension* around 1920.

Tandem and reversing cold mills for the production of sheet and tinplate products have been widely used. However, it must be realized that hot rolling must still be used in reduction an *ingot* to a strip with a thickness on *the order of* 6mm. Attempts to cold roll strip of greater thickness would necessitate excessive rolling forces and energy requirements.

3.6 Modern Cold-reduction Facilities

In the 1930's, the cold reduction of hot-rolled steel strip evolved from a rather specialized, small-scale process to a position of prime importance in the production of cold-rolled sheets and strips. These products differ from each other principally in dimensions.

The maximum available widths of cold-rolled strip increased rapidly from 1925 onward and the minimum thickness for a given width decreased. By 1937, the thickness-width limits for both hot- and cold-rolled strip had reached greater values, and today *flat-rolled* products are available in even greater width-to-thickness ratios than before.

Typical of the sheet mills built in the 1930's is the three-stand, 2130mm tandem mill. This mill utilized 520mm diameter work rolls, 1420mm diameter backup rolls, and was driven by motors totaling 5100kW at speeds up to 165m/min. Just before and subsequent to World War II, four-stand sheet mills came into vogue, but in the 1960's five-stand sheet mills were built.

For tinplate production (with final thicknesses of rolled strip on the order of 0.3mm) *five-stand tandem mills* soon came into use in the 1930's. Such a mill installed at the Irvin Works in 1941 used 530mm and 1350mm by 1220mm rolls and possessed a total motor drive of 8280kW. It could deliver strip at speeds up to 1140m/min. Later mills for tinplate production utilized larger motors and were generally operated at speeds up to 1520m/min. Six-stand tin mills were introduced in the 1960's with still more installed horsepower and slightly larger workrolls. They are operated partially under the control of computers.

Although tandem cold mills constructed in recent years have utilized conventional 4-high mill stands, a unique tandem mill facility specifically designed for rolling *stainless sheet* products up to 1270mm wide was put into service in 1969 by Nisshin Steel Company at the Shunan Works located in Nanyo, Japan. This mill was the first mill designed for *fully continuous* operation with the incoming coils fed from *pay-off reels* through a *welder* to be jointed head-to-tail. It utilized a train of 6 stands, the first and last stands being 2-high mills and the *intermediate stands* being *Sendzimir mills*.

Another fully continuous cold mill for rolling steel strip was put into operation in 1971 at Nippon Kokan's Fukuyama Works in Japan. This mill also featured two pay-off reels, a *shear*, welder, *strip accumulator*, five mill stands, a *flying shear* and two *tension reels*. Coils to be rolled are welded end to end and, during the welding operation, strip continues to be drawn from the accumulator. The mill utilized 4-high stands, as fully automated as present technology permits and is under computer control.

For stainless and *silicon steel* rolling and for the processing of *special alloys* with limited markets, *single-stand reversing mills* have continued to find popularity. Such mills usually feature powerful main drives and motors attached to the reels supplying power approximately equal to half that of the main drive motor.

The foregoing mills are operated differently from their early slow predecessors in a number of ways. High strip tensions are now usually used and *rolling lubricants* were found to be necessary. Such lubricants were found to greatly facilitate the rolling operation in reducing rolling forces and lessening the rolling energy required. An aqueous recirculated mixture of the rolling lubricants was used not only to provide the lubricity in the *roll bite* between the roll and strip surfaces, but also acted as a coolant for the rolls and the strip being rolled.

Mill speeds have also increased considerably but coincident with this has been a steady improvement in rolled strip quality. Automatic gage control systems have maintained good uniformity in the thickness of the rolled product and improved lubricants have maintained strip coiling temperatures within permissible limits.

3.7 Foil Mills

In recent decades, various metals have been rolled to very *light gages* for such application as *packaging*, *capacitor* manufacture, and *printed circuit boards*. For virtually all these applications, however, the quantity of rolled product was limited. Accordingly, small mills were used in making such specialized foil in thickness often in the range 2.54×10^{-3} mm to 2.54×10^{-2} mm.

Soon after 1960, however, it became apparent that a reasonably large market existed for relatively wide (up to 760mm or more) steel foils. A foil mill capable of rolling tin coated strips down to a thickness of 3.81×10^{-2} mm and less in widths up to 760mm or more was therefore installed at the Gary Works of the U. S. Steel Corporation in 1965.

3.8 Temper or Skin Pass Mills

Basically such mills, give a very light reduction to *annealed strip* so as to provide a degree of surface hardening, prevent the breaking of the surface in subsequent drawing operation. In addition, temper mills are used to impart a designed finish or luster surfaces of the workpieces and are frequently used to impart the designed degree of *flatness* to the rolled product.

Usually temper mills are operated without a rolling lubricant. This is fortunate in that the rolled surfaces of the strip remain virtually *uncontaminated* and, therefore, ready for further processing (such as *tinning*). Moreover, the high *friction* occurring in the roll bite between the roll and strip surfaces ensures that only a very limited elongation or reduction is given to the workpiece.

Occasionally, however, temper mills are operated "wet" either for the purpose of achieving a larger reduction than would otherwise be obtainable and/or with the intent of leaving a *corrosion-resistant* or other type of film on the surface of the temper rolled material.

Temper mills for the rolling of sheet products have traditionally utilized a single-stand while two-stand mills are commonly used to produce the harder tempers required for tinplate products.

In recent years, there has been a tendency to utilize one rolling facility for two or more different types of rolling operations. For example, single and two-stand mills have been installed to carry out both cold reduction and temper rolling operations. In such facilities, provision is usually made to accommodate sets of work rolls of different sizes, smaller diameter rolls being used for normal cold rolling operation and larger rolls for temper rolling.

3.9 Modern Hot Strip Mills

Although the basic concepts associated with hot rolling have not, in general, changed appreciably since the last century, mills have become considerably larger, faster, more powerful and capable of rolling products of larger sizes to *closer dimensional tolerance* and improved surface finishes. Contributing to this situation are improved *mill components* including rolls, housing, drive motors, *instrumentation* and *control systems*, which have resulted from the impressive advances made in the *engineering science* during the 20th century. Paralleling these advances has been a better theoretical understanding of *rolling technology* developed in various industrial and governmental research laboratories throughout the world. In addition, it should be noted that hot rolling is not only used for *shaping* purposes but is now used to obtain the designed *metallurgical properties* in the rolled products (controlled rolling), for the *cladding* of various metals to steel and to impart desired surface finishes particularly *in the case of* flat-rolling materials.

With respect of flat-rolling equipment, the *conventional four-high mill stand* with its d-c motor drive has continued to demonstrate the soundness of its basic design in spite of the advent of mills of unconventional design, such as *planetary* and *six-high mills*. Though the arrangement of the rolls in a four-high mill has remained unchanged, such mills have increased in physical size and drive power. For example, plate mill widths have increased to 5500mm, work-roll diameter up to 1220mm, back-up roll diameter up to 2390mm and drive power up to 1790kW with the electrical power supplied nowadays by *silicon-controlled rectifiers* instead of motor-generator sets. In addition, modern tandem-rolling-mill trains utilize an increased number of mill stands.

The conventional mill housing is, however, slowly being changed. *Screws* are being gradually replaced by *hydraulic roll-positioning system* which respond much more rapidly than screws to *automatic-gage-control* (AGC) signals and provide adjustable mill *stiffness*. Moreover, *roll-bending* and *roll-bending jacks* are now being incorporated in the mill housing rather than in the individual roll chocks.

With the availability of electronic computers, the control of rolling mill has become more and more *sophisticated*. Programmed with *mathematical models* of the rolling process, computers are now being used to control the *pacing* and *set-up* of hot mills and the *cross-sectional profile* and *shape* of hot-rolled strip.

In recent years the direct rolling of hot workpieces, such as *slabs* has been practiced as a method of conserving energy and increasing yield.

A further development along these lines has been the *in-line* rolling of *continuously cast billet* and *slab strands*. Nucor Corporation has built a system linking a *billet casting machine* to a *bar mill* in which the rolling speed is varied to match the throughout of the caster and a similar system connecting a *rotary caster* to a billet mill has been developed in Japan.

Around mid-century, the typical hot-strip mill then being built utilized two or three *furnaces*, a *scale-breaker*, four or five *roughing stands*, a second scale-breaker, a *six-stand finishing train* and two *coilers* at the end of a *run-out table*. Such mills could produce from about 1.5 to 2.5 mi-llion tons of hot-rolled strip per year. To meet increased production requirements, the "*Generation II*" hot strip mill of the 1960's, as exemplified by U. S. Steel's 2130mm facility at its Gary Works, features four slab-reheat furnaces of the five-zone type, a *vertical edger*, a horizontal scale-breaker, three individual roughers, a *two-stand tandem-roughing mill*, a finish scale-breaker, seven *finishing stands* and three coilers. The *main drives* associated with the mill are rated at 92100kW and the *auxiliary drive* and other equipment are rated at an additional

55550kW. This facility has a capacity in excess of 3.5 million tons per year.

Not only is the hot-strip mill of today more productive, it is virtually under complete computer control. Slabs may be *charged and discharged* from the reheat furnaces as dictated by the computer and the finishing train, the *cooling system* on the run-out table and the coilers are all automatically controlled to provide the desired finishing and coiling temperatures.

In the operation of such mills, a number of distinct advances have been made. The *cast-iron and steel work rolls* originally used are giving way to rolls of improved resistance to wear and *fire-cracking*, such as *high chromium and centrifugally cast rolls*. Moreover, rolling lubricants are becoming more extensively used on the finishing trains of hot-strip mills to minimize *rolling force* and power requirements, to extend the life of the mill rolls and to enhance the surface quality of the rolled strip. In addition, many mills feature either work or *back-roll-bending systems* to control the flatness of the rolled strip. Furthermore, during the last decade, cooling of the rolled strip on the run-out table has been accomplished more efficiently by the *so-called "laminar-flow"* low-pressure technique rather than by conventional high-pressure water sprays.

An increasing innovation incorporated in a new hot-strip mill recently commissioned by John Lysaght (Australia) Ltd. is a *coil box* located after the last rougher for temporarily storing the partially rolled bar in coiled form prior to finishing. This feature enables the length of the *holding table* ahead of the finishing train to be decreased in length, thereby reducing the overall length of the mill and the capital cost associated with its installation.

Controlled rolling on the plate mill involves rolling at lower-than-normal temperature so that a smaller grain size at ambient temperature is achieved in the rolled *microstructure*. As a consequence of the smaller grain size, the *yield strength* and the *toughness* of the rolled product are usually significantly improved. However, these qualities are generally attained only with higher rolling forces, delay in processing so as to achieve lower temperatures for the final pass and greater difficulties in obtaining flatness and poper cross-sectional profiles in the rolled workpieces.

In the rolling of *structure steel*, there has been a trend towards increased lengths of workpieces through the use of continuous and semi-continuous mills and the use of subsequence *cold sawing*. *Cooling beds* have been built to accept beams 6350mm long and highly automated *warehousing* is practiced.

Bar mills have been generally resisted attempts at supplicated control and precise *measuring devices* were found to be unreliable due to the harsh environment. However, there have been considerable mechanical improvements to bar mills with better-controlled, higher-speed drives. A completely new concept in bar rolling has been developed by Schloemann-Siemag. This is the *three-roll planetary mill* designed to make *large reductions* (up to 92%) in a *single pass*. Another recent innovation in bar mill design has been the use of two 3-high roughing stands in tandem with individual gear drives coupled to non-reversing a-c motors.

Rod mills have been built with ever-increasing speeds until today they have attained 20,000 fpm. Since the introduction of the *no-twist mill* and *Stelmor Process lines* in the early 1960's, over 140 of the former and 150 of the latter are in worldwide operation.

In the production of *seamless tubes*, the current technology is still based on *Mannesmann type piercing*, *Assel*, and *stretch-reducing mills*.

Words and Expressions

rolling 轧制，轧钢
ductility 塑性，韧性，可锻性
cold 冷加工
goldsmith 金匠
jewelry 珠宝
works of art 手工艺品
roll 轧辊
gold, silver, lead 金、银、铅
rolling mill 轧机
worm gear 蜗轮
lead sheet 铅板（片）
tapered lead bar 楔形铅块
embodying 采用
thickness 厚度
mint flat 造币板
mint 造币厂
patent 专利（权）
slitting mill 纵切机
disc 圆盘
spindle 轴
flat bar 扁平轧件
revolving 旋转
strip 带钢
roofing 屋顶
flashing 防水
tin 锡，马口铁
pipe 钢管，管道
strong-armed cross 强力十字手柄
axle 轴
with the exception of 除…以外
pertain to 适合
softer metals 软金属
presumably 大概
birthplace 诞生地

claim 声称，主张
priority 优先（权）
led the way 领先
thin flat 薄板
slitting 纵切（分）
foreshadow 预示
fruition 实现，成就
finishing 修整，精整
plane-surfaced rolls 平辊
flatten out 轧（碾）平
irregularities 不规则物体
pamphlet 小册子
hammer 铁锤，锤子
large rolling mills 大型轧机
hot rolling 热轧
ferrous materials 黑色（金属）材料
rod 棒（材）
ironworks 制铁（钢）厂
plate （金属）板
monopolize 独占，垄断
tinplate 镀锡板
terne plate 镀铅板
Lauth mill （三辊）劳特轧机
backup roll 支持辊
equipped with 装备
roller table 辊道
of the day 当代
reversing mill 可逆轧机
clutch 离合器
gear system 齿轮系统
plain roll 平辊
grooved roll 带槽轧辊
cast lead ingot 铸造铅锭
mandrel 芯轴

Words and Expressions

steelmaker 钢铁工人
cold rolling 冷轧
in vogue 流行
grant 授予
coupling box 联结箱
nut pinion 人字齿轮
in unison 一致
rates of revolution 转速
wear 磨损
guide 导卫（板）
hand-forged plate 人工铸造板
elongation 延伸
pack rolling 叠轧
ply rolling 叠层轧制
single-stand 单机架
two roll 两辊
hot mill 热轧机
piece 工件，轧件
top roll 上辊
hand tong 手钳
cooled down 冷却
reheat 再加热
bare outline 简单总结
owing to 由于
conventional sheet hot mill 常规板带热轧机
sheet mill 薄板轧机
two-high, pull-over mill 两辊往返式轧机
rolling unit 轧制机组
tandem mill 连轧机
successive stand 串列机架
wire rod 线棒（材）
four-high tandem mill 四辊连轧机
mechanical guide 机械导卫
rod mill 棒材轧机
tilt hammer 锻锤
successor 继承者，接任者
contest 争论
industrial revolution 工业革命
momentum 动力

iron and steel 钢铁
rail rolling mill 轨梁轧机
wrought iron 锻铁，熟铁
I-beam 工字梁
the British Great Exposition 英国展览会
three-high mill 三辊轧机
heavy section 大型型钢
housing （轧机）机架
upper roll 上辊
lower roll 下辊
lifting table 升降辊道
middle roll 中间轧辊
reversing plate mill 可逆中板轧机
steamship 轮船
universal mill 万能轧机
continuous hot mill 连续式热轧机
rod rolling mill 棒材轧机
billet 小方坯
in line 一致，协调
horizontal 水平的
vertical 垂直的
predecessor 前任
cluster mill 多辊轧机
work roll 工作辊
Z-bars Z型轧件
beam 钢梁
H-beam H型钢梁
upsurge 高潮
semi-continuous hot strip mill
　　半连续热带轧机
two-high tandem finishing train
　　两辊连轧精轧机组
width 宽度
thickness 厚度
length 长度
roughing train 粗轧机
two three-high stands 两架三辊轧机
finishing train 精轧机组
five-stand 五机架

commercial   商业上的
to all intents and purposes   实际上
metalworker   冶金工人
the British Parliament   英国国会
disregard   忽视
pig iron   生铁
hammered bar   锻件
forge   铸造
nail rod   制钉圆棒
pass   道次
waterwheel   水轮
chilled iron roll   冷硬铸铁轧辊
roll neck   辊颈
the American Revolution   美国独立战争
boiler plate   锅炉板
steel corporation   钢铁公司
incidentally   附带地，顺便地
angle iron   角钢
foster   培养，鼓励
inventiveness   独创性
corrugated plate   波形板
mill housing   轧机机架
cast guide rails   铸造导轨
U. S. rights   美国版权
the Civil War   （美国）国内战争
stimulus   刺激，促进因素
iron and steel industry   钢铁工业
blooming mill   初轧机，方坯初轧机
reversing blooming mill   可逆初轧机
two-high reversing blooming mill
   两辊可逆初轧机
three-high roughing mill   三辊粗轧机
three-high finishing mill   三辊精轧机
horsepower   马力
reservoir   蓄水池，水库
shaft   轴
coupling   联结器
flywheel   飞轮
reversal   可逆的

electric generator   发电机
internal combustion engine   内燃机
gas-driven   蒸汽驱动
reciprocating steam-driven   往复式蒸汽驱动
direct current motor   直流电机
main drive   主传动（系统）
multi-armature motor   多电枢电机
cold reduction facilities   （金属）冷加工设备
reduction   压下
steel band   钢带
high-carbon   高碳
impetus   推动力，促进
Lauth 3-high cold mill   三辊劳特冷轧机
wrought-iron work rolls   锻造工作辊
cast-iron backing roll   铸造支持辊
cold rolled strip   冷轧带钢
2-high mill   两辊轧机
roll neck bearing   辊颈轴承
reversing cold mill   可逆冷轧机
tandem cold rolling   冷连轧机
2-high 4-stand tandem mill   两辊四机架连轧机
tension   张力
tension reel   张力卷取（机）
synchronize   同步
cold strip reel   冷轧带钢卷取机
hot strip reel   热轧带钢卷取机
high-tension   大张力
constant tension   恒张力
current control   电流控制
concurrent   同时发生的，一致的
strip tension control   带钢张力控制
slack strip   松弛带钢
loop take-up   活套张紧器
interstand tension   机架间张力
ingot   钢锭
the order of   …数量级
flat-rolled   扁平轧制的
five-stand tandem mill   五机架连轧机
stainless sheet   不锈钢板

fully continuous 全连续
pay-off reel 开卷机
welder 焊接机
intermediate stand 中间机架
Sendzimir mill 森吉米尔轧机
shear 剪切机
strip accumulator 带钢收集器
flying shear 飞剪
tension reel 张力卷取机
silicon steel 硅钢
special alloy 特殊钢
single-stand reversing mill
　单机架可逆轧机
rolling lubricant 轧制润滑
roll bite 轧制变形区
foil mill 箔材轧机
light gage 薄规格（厚度）
packaging 包装
capacitor 电容器
printed circuit board 印刷电路板
temper or skin pass mill 平整机
annealed strip 退火带钢
flatness 平直度（板形）
uncontaminated 无污染的
tinning 镀锡
friction 摩擦
corrosion-resistant 抗腐蚀
closer dimensional tolerance
　更精确的尺寸公差
mill components 轧机部件
instrumentation 仪表
control system 控制系统
engineering science 工程科学
rolling technology 轧制技术
shaping 成型
metallurgical properties 冶金性能
cladding 复合
in the case of …情况下
with respect of 关于

conventional four-high mill stand
　常规四辊轧机
planetary mill 行星轧机
six-high mill 六辊轧机
silicon-controlled rectifiers
　可控硅整流器
screw 压下（螺丝）
hydraulic roll-positioning system
　液压轧辊定位系统（液压APC）
automatic-gage-control（AGC）
　厚度自动控制系统
sophisticated 复杂的
mathematical model 数学模型
stiffness 刚度
roll-bending 轧辊弯曲
roll-bending jacks 轧辊弯曲装置
pacing 节奏控制
set-up 轧机
cross-sectional profile 横断面形状
shape 板形
slab 板坯
in-line 在线
continuously cast billet 连铸小方坯
slab strand （连铸）扁坯
billet casting machine 小方坯连铸机
bar mill 小型轧机
rotary caster 旋转式连铸机
furnace 加热炉
scale-breaker 除鳞机
roughing stand 粗轧机座
six-stand finishing train 六机架精轧机组
coiler 卷取机
run-out table 输出辊道
"Generation Ⅱ" hot strip mill
　第二代带钢热轧机
vertical edger 立辊轧机
two-stand tandem-roughing mill
　两机座粗轧连轧机
finishing stand 精轧机座

main drive　主传动
auxiliary drive　辅助传动
charged and discharged　装钢、出钢
cooling system　冷却系统
cast-iron and steel work roll
　铸铁/铸钢工作辊
firecracking　裂纹
high chromium and centrifugally cast rolls
　高铬离心铸造轧辊
rolling force　轧制压力
back-roll-bending system
　支持辊弯曲系统
so-called　所谓的
laminar-flow　层流冷却
coil box　卷取箱
holding table　中间辊道
microstructure　微观组织

yield strength　屈服强度
toughness　韧性
structure steel　结构钢
cold sawing　冷锯
cooling bed　冷床
warehousing　仓库
measuring device　测量装置
three-roll planetary mill　三辊行星轧机
large reduction　大压下量
single pass　单道次
fpm　feet per minute
no-twist mill　无扭轧机
Stelmor Process lines　斯泰尔莫线
seamless tube　无缝钢管
Mannesmann type piercing　曼内斯曼穿孔
Assel　阿塞尔轧管
stretch-reducing mills　张力减径机

4 Classifications of Rolling Mills

4.1 Main Components of a Mill Stand

A typical mill for rolling of flat products includes one or a number of the *mill stands* which are arranged, usually in line, to produce a sequential *reduction* in thickness and width of the rolled product. Although a variety of the mill stand designs are known, it is possible to identify their common *components* from functional viewpoint.

Fig. 4-1 and Fig. 4-2 illustrate schematically a mill stand designed for reducing the workpiece thickness. Main functional components of this mill stand are:

(1) *Work rolls* between which the rolled product is being squeezed.

(2) *Backup rolls* which support the work rolls to reduce their *deflection* under rolling load.

(3) *Roll gap adjustment mechanisms* provide setting of required gap between work rolls and may also allow one to adjust elevation of the pass line.

(4) *Housing* is designed to contain the mill stand components and to withstand the rolling load.

(5) *Main drive train* provides rotation of the rolls with desired speed and *rolling torque*.

Generally, the same main components can be found in a typical mill stand designed for reducing workpiece width.

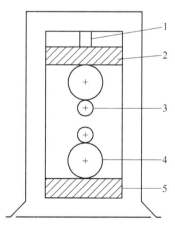

Fig. 4-1 Schematic illustration of a horizontal mill stands
1—roll gap adjustment cylinder; 2—housing; 3—work roll;
4—backup roll; 5—roll bending cylinder

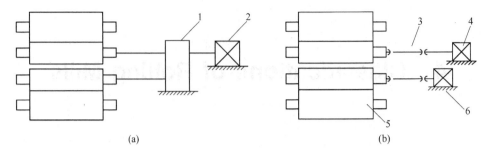

Fig. 4-2 Schematic illustration of two horizontal mill stand
1—pinion stand; 2—motor; 3—spindle; 4—top motor; 5—roll mill; 6—bottom motor

4.2 Classification of Mill Stands

Mill stands for rolling flat products can be classified by the following categories:

(1) *Roll arrangement*—Depending on the roll arrangement in the *mill housing*, the mill stands are referred to as:

1) *Two-high mill stand*. Two-high mill stand contains two work rolls (Fig. 4-3(a)).

2) *Three-high mill stand*. In three-high mill stands (Fig. 4-3(b)), the *top and bottom rolls* revolve in the same direction and the *middle roll* in the opposite direction. After completing the bottom pass (solid line), the workpiece is lifted to provide the reversing top pass (dotted line).

3) *Four-high mill stand*. This is the most common roll arrangement that includes two *work rolls* and two *backup rolls* (Fig. 4-3(c)).

4) *Five-high mill stand*. In the five-high mill stand arrangement, an intermediate roll is added between one of the work rolls and a backup roll (Fig. 4-3(d)).

5) *Six-high mill stand*. In addition to two work rolls and two backup rolls, the six-high mill stand has two intermediate rolls (Fig. 4-3(e)).

6) *Cluster type mill stand*. Main feature of the cluster type mill stand is that each work roll is surrounded by more than one *intermediate roll* which in their turn is supported by a number of backup rolls. Both *asymmetrical* (Fig. 4-4(a)) and *symmetrical* (Fig. 4-4(b)) cluster type rolling mill stands have been developed.

7) *Mill stand with off-set rolls*. In this mill stand arrangement, at least one of the work rolls is displaced to one side from the vertical centerline of the backup rolls. The displaced roll is laterally supported by intermediate rolls (Fig. 4-5).

(2) Direction of *roll axes*—The mill stands can be identified in association with direction of the roll axes as following:

1) *Horizontal mill stand*. In the horizontal mill stand all rolls are parallel to the mill floor (Fig. 4-6(a)).

2) *Vertical mill stand*. The roll axes of vertical mill stands are perpendicular to the mill floor (Fig. 4-6(b)).

4.2 Classification of Mill Stands

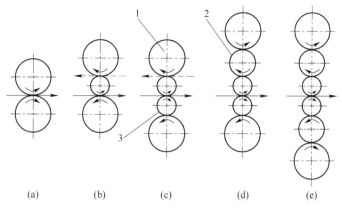

Fig. 4-3 Types of roll arrangements in the mill stands
1—backup roll; 2—intermediate roll; 3—work roll

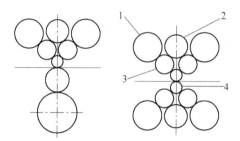

Fig. 4-4 Cluster type of mill stand
1—side backup roll; 2—center backup roll; 3—intermidiate roll; 4—work roll

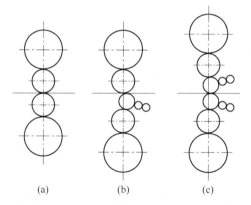

Fig. 4-5 Types of mill stands with offset rolls
(a) 4-high mill stand; (b) 5-high mill stand; (c) 6-high mill stand

3) *Crossed-roll mill stand*. In the crossed-roll mill, the axes of the rolls are tilted in opposite directions in respect to rolling direction (Fig. 4-6(c)).

4) Mill stand with parallel-tilted rolls. In this type of mill stand, the rolls are tilted at the same angle in respect to rolling direction (Fig. 4-6(d)).

(3) Direction of rolling—Three types of the mill stands are usually considered with regard to

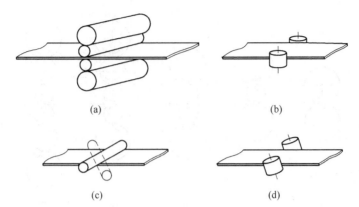

Fig. 4-6 Mill stands with different directions of the roll axes
(a) horizontal mill stand; (b) vertical mill stand; (c) crossed-roll mill stand; (d) mill stand with parallel-tilted rolls

rolling direction:

1) *Non-reversing mill stand*. This mill stand is designed to roll in one direction.

2) *Reversing mill stand*. This mill stand provides rolling in both directions.

3) *Back-pass mill stand*. In the back-pass mill stand, the rolls are always rotating in the same direction. After completing the rolling pass, the roll gap is opened and the workpiece is back-passed toward the entry side of the mill. Then the roll gap closes and the next rolling pass proceeds.

(4) Main motor type—There are two principal types of main drive motors which are used in rolling mill stands:

1) A-C motors. The alternative current (A-C) motors can be of three modifications: synchronous, squirrel cage and wound-rotor induction motors. They are generally used to provide rotation of the rolls in the same direction at practically constant speed.

2) D-C motors. The direct current (D-C) motors provide reverse rotation of the rolls with wide range of the roll speed control.

(5) *Drive train* arrangement—The principal drive train arrangements can be classified as follows:

1) *Direct drive*. The direct drive train provides connection of the motor with driven rolls without any change in angular speed.

2) *Gear drive*. In the gear drive, the angular speed of the motor is either reduced or increased by installing a *gear box* in the drive train.

3) *Pinion stand drive*. The drive train with a *pinion stand* (Fig. 4-2(a)) allows one to drive both top and bottom rolls from a single motor.

4) *Independent drive*. In the independent drive train (Fig. 4-2(b)), top and bottom rolls are driven by independent motors.

The drive train arrangements can also be identified in relation to what type of roll is driven:

1) Train with driven work rolls.

2) Train with driven backup rolls.

3) Train with driven intermediate roll.

(6) Special design mill stands—A number of mill stands of unconventional design have been developed. They include:

1) *Planetary mills.*
2) *Rolling-Drawing mills.*
3) *Reciprocating mills*, etc.

4.3 General Classification of Rolling Mills

In general terms, rolling mills can be classified in respect to rolling temperature, type of rolled product, and type of the mill stand arrangement.

(1) Rolling temperature—In regard to rolling temperature, the rolling mills are usually identified as:

1) Hot rolling mills. Hot rolling process of steels generally begins when the workpiece temperature is equal or less than 1315℃ and completes at the temperatures which are either above or slightly lower than A_3 *critical temperature* which, for low-carbon steel, is about 900℃. Thus, the bulk of the rolling process occurs when the rolled material is in *austenitic phase*.

2) Cold rolling mills. Cold rolling process is usually conducted with a workpiece that has initial temperature equal to a room temperature. During cold rolling the rolled material temperature may rise to between 50℃ and 65℃.

3) Warm rolling mill. *Warm rolling process* is generally conducted at elevated temperatures substantially lower than A_1 *critical temperature* which, for low-carbon steel, is approximately 730℃.

(2) Rolled product—In relation to a type of rolled product, the rolling mills are identified as:

1) *Slabbing mills.* The slabbing mills roll *ingots* into slabs which usually vary in thickness from 150mm to 300mm. The mill stand may have a provision for a large opening of the roll gap, so the width reduction can be made by rolling the slab on its edge.

2) *Plate mills.* The hot plate mills roll slabs into plates. The rolled product can be either in a flat form or in a coil form with a consequent uncoiling and cutting the product into the plates of desired lengths. The most distinguished characteristic of both hot and cold plate mills is the width of the rolled plates which for some mills may be as wide as 5334mm.

3) *Strip mills.* Conventional hot strip mills reduce a slab to a strip with thickness as thin as 1.2mm. Cold strip mill further reduce the strip to a desired final gauge. The width of the steel coils is usually between 600mm and 2000mm.

(3) Mill stand arrangement—Depending on the distance between the two adjacent mill stands in relation to the length of the rolled product, the mill stand arrangements are usually referred to as:

1) Open mill stand arrangement. In open mill stand arrangement, the distance between two adjacent stands is always greater than the length of the rolled bar exiting the upstream mill stand. This allows rolling on these stands with independent speeds.

2) Close-coupled mill stand arrangement. In the close-coupled mill stand arrangement, the distance between adjacent stands is less than the length of the rolled bar exiting the upstream mill

stand. So the bar is rolled simultaneously on adjacent stands. This requires synchronization of speed of these mill stands to provide a constant mass flow of the metal.

The close-coupled mill stand arrangements are used in two types of well-known rolling mills:

①*Universal rolling mill*. The universal rolling mill includes a *horizontal mill stand* and, at least, *one vertical mill stand*. In this mill both edging and flat rolling may be conducted simultaneously.

②*Tandem rolling mill*. In tandem rolling mills, two or more mill stands of the same type (usually horizontal) have a close-coupled arrangement with each other.

4.4 Components of High-production Hot Strip Mills

The term "high-production hot strip mill" is conventionally applied to the mills with yearly production rate equal to or greater than 2 million metric tons. The main components of these mills (Fig. 4-7) are briefly discussed below:

(1) *Reheat furnaces*—In the reheat furnaces, cold or warm slabs are heated to a desired temperature for rolling.

(2) *Roughing mill*—The roughing mill, or train, consists of a series of horizontal and vertical mill stands. The main purpose of the roughing mill is to roll a slab into a *transfer bar* with thickness which is commonly between 19mm and 45mm.

(3) *Transfer table*—The transfer table, or *delay table*, is located between roughing and *finishing mill*. The table is usually long enough to accommodate a full length of the transfer bar. This allows one to roll at least two bars independently, one on the roughing mill and another on the finishing mill.

(4) *Shear*—The shear is located in front of finishing mill. It is usually designed to cut both the head and tail ends of the transfer bar prior to their entry into the finishing mill.

(5) Finishing mill—The finishing mill, or train, consists of one or a series of horizontal mill stands. In *tandem finishing train*, there are *loopers* installed between stands. Each looper maintains a desired *interstand strip tension* by pushing a free rotating roller against the strip.

(6) *Runout table*—The runout table is located between the finishing mill and coilers. A series of water *cooling headers* are installed above and under runout table. The water coolant system is designed to reduce the strip or plate temperature before the rolled material enters a coffer.

(7) *Coilers*—The coilers are usually located at the end of the runout table. In some cases, when cooling of the strip is not required, the coilers may be installed right after the finishing mill.

(8) *Descaling system*—Removing the scale from the surfaces of the rolled piece is provided by using a series of high-pressure water spray headers installed at different location of hot strip mills. In some hot strip mills, vertical and horizontal mill stands, known as *scale breakers*, are added to improve efficiency of descaling process.

(9) *Roll coolant system*—In the mill stands, the rolls are cooled with the water spray headers located in close vicinity with the rolls. In some finishing mill stands, the roll coolant system is supplemented with the roll lubrication system.

(10) *Interstand cooling system*—The interstand cooling system is installed in some high-speed finishing trains to reduce the strip temperature. The strip is cooled by the water sprays located between the mill stands.

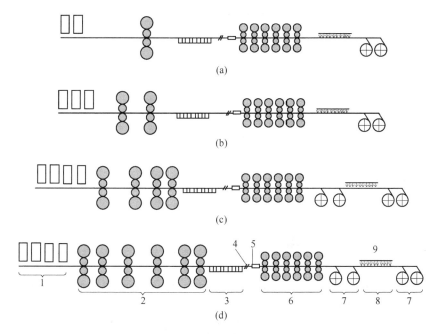

Fig. 4-7 Principal layouts of hot strip mills

(a) *semi-continuous*; (b) with *twin reversing roughing train*; (c) *three-quarter continuous*; (d) *fully continuous*

1—furnace; 2—roughing mill; 3—transfer table; 4—shear; 5—scale breaker;
6—finishing mill; 7—coilers; 8—run-out table; 9—water coolant headers

4.5 Classification of High-production Hot Strip Mills

Finishing trains of the high-production hot strip mills have customarily from four to seven horizontal mill stands arranged in tandem. The principal difference between hot strip mills is mainly in layouts of the roughing mills. This difference is often used as a base for their classification into four distinct types.

(1) Semi-continuous hot strip mill—The roughing mill of the semi-continuous hot strip mill has one vertical and one horizontal reversing mill stand which are usually combined into one *universal roughing mill stand* (Fig. 4-7(a)). These mills may also include both vertical and horizontal *scale breaker* installed upstream in relation to the reversing rougher.

(2) Hot strip mill with twin reversing roughing train that contains two universal roughing mill stands (Fig. 4-7(b)).

(3) Three-quarter continuous hot strip mill—The roughing train of the three-quarter continuous hot strip mill has one or more horizontal *single-pass stands* after reversing roughing mill, in open or close-coupled configuration (Fig. 4-7(c)).

(4) Fully continuous hot strip mill—Fully continuous hot strip mill has four or more horizontal roughing mill stands, the last two stands being either open or close-coupled.

The mills of different generations can be recognized not only by their layouts but also by their production rate and the weight of the rolled coils.

Generation I : 1927~1960

Yearly production rate 1.0 to 3.0Mt

Coil weight per unit width 4.0 to 11.0kg/mm

Generation II : 1961~1969

Yearly production rate over 3.0Mt

Coil weight per unit width 18.0 to 22.0kg/mm

Generation III : 1970~1978

Yearly production rate over 5.0Mt

Coil weight per unit width up to 36.0kg/mm

4.6 Compact Hot Strip Mills

By definition, the compact hot strip mills are shorter and comprise lesser number of the mill stands. Fig. 4-8 gives a diagrammatic comparison of the layouts of some compact hot strip mills with a layout of semi-continuous hot strip mill (Fig. 4-8(a)).

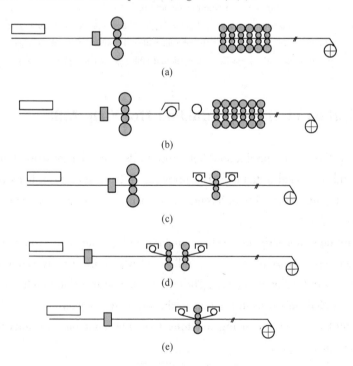

Fig. 4-8 Comparison of layouts of compact hot strip mills with a layout of semi-continuous hot strip mill

Coilbox arrangement—The length of the transfer table between roughing and finishing mills can be reduced by installation of a coilbox in front of the finishing train (Fig. 4-8(b)). After completing the last roughing pass the transfer bar is coiled in the coilbox; then the coil is transferred downstream and uncoiled prior to entering the finishing train.

Reversing finishing mill arrangement—This type of mill comprises a reversing roughing mill and a reversing finishing mill known as a *Steckel mill*. In this mill, to conserve heat during intermediate finishing passes, the roiled strip is coiled on the preheated drums adjacent to the mill stand. The length of the hot strip mill with the Steckel mill can be further reduced by using the close-coupled arrangement between roughing and finishing mills as shown in Fig. 4-8(c).

Twin reversing hot strip mill—In this arrangement, the distance between reversing roughing and finishing mill stands is very short and the coiling furnaces are located as shown in Fig. 4-8 (d). Roughing passes are intended to be rolled without coiling. Since the mill stands are close-coupled, the rolling passes can be made simultaneously on both stands.

Single stand reversing hot strip mill—This mill is designed to roll both roughing and finishing passes using the same mill stand (Fig. 4-8(e)). This is the simplest hot strip mill arrangement which is used for rolling both strip and plate in coil form.

4.7 Integrated Continuous Casting and Hot Rolling Process

Integration of continuous casting process with hot rolling process allows one to eliminate reheating of slabs and thus to conserve energy, improve quality and yield.

Integration with thick continuous caster—*Hot Direct charging Rolling* (*HDR*) combined with continuous casting machine developed by Nippon Kokan K K to achieve desired temperature of the bar prior to its entry into roughing mill, the following equipment is added:

(1) Induction type *edge heater* that provides uniform bar temperature.

(2) *Heat insulator* that reduces heat losses of the bar during its transfer toward hot strip mill.

(3) An additional heat insulator on transfer table between roughing and finishing mill.

Integration with thin continuous caster—When thin continuous caster is used, both the slab reheating furnace and the roughing mill are no longer needed for hot strip mill operation.

In the concept developed by SMS Schloemann—Siemag AG, the continuously cast thin slab is fed directly into four-stand finishing mill. Prior to entering the mill, the slab is cut with a shear and then is passed through a *reheating tunnel furnace*. The main parameters of the facilites are listed in Table 4-1.

Table 4-1 Main parameters of the facilities (mm)

Main parameters of the facilities	Concast slab dimensions
thickness	40~50
width	1200~1600
casting rate	5.0~6.0
rolled strip thickness	1.5~25

In the concept proposed by Kawasaki Steel, two thin casters are used to feed one finishing hot strip mill with four six-high mill stands. After reheating, the *concast* slabs are coiled and are then held in a coil holding furnace. When finishing mill is ready to roll, the uncoiled thin slab passes through an edge heater before entering the mill.

The *thin continuous caster* has also been proposed to be integrated with a planetary mill and with a Steckel mill. After passing through the *equalizing furnace*, the continuously cast thin slab is rolled at the planetary mill. The thin strip is then rolled at the *planishing mill* to smooth the strip surface.

4.8 Cold Mill Arrangement

The most common cold mill arrangements are:

(1) *Single-stand cold mill*—Two types of the single-stand cold mills are known:

1) *High-reduction* cold mills which are capable of taking up to 50% reduction per one pass;

2) *Skin-pass*, or *temper mills*, which are designed for reductions ranging from 0.5% to 4.0%. Single-stand mills can be either reversing or non-reversing.

(2) *Twin-stand*, or *double-reduction cold mill*—This is commonly a *non-reversing mill*.

(3) *Tandem cold mill*, Three or more close-coupled non-reversing mill stands usually constitute a tandem cold mill. The tandem cold mills can be divided into two groups: stand alone and fully continuous cold mills.

1) Stand-alone cold mill. The coil to be rolled is delivered to the entry side of the mill by the *entry coil conveyor* and is then lifted to uncoiling position by an *entry coil lifting car*. Between mill stands and sometimes after last mill stand, there are free-rotating rollers which are used as the parts of the *strip tension* measuring devices. The strip exiting the last mill stand is coiled on a *tension reel*. After coiling is completed the exit coil lifting car removes the coil from the tension reel and transfers it on to the exit coil conveyor.

2) Fully continuous cold mill. In these facilities (Fig. 4-9), the *provisions* are made for welding the strips head to tail to provide continuity of rolling during welding; the excess of strips is stored prior to welding operation and is released during welding operation.

4.9 Processing Lines Incorporating Cold Mills

The following steps are typically involved in processing of the coils after hot rolling:

(1) Descaling—Main purpose of descaling is to remove the scale accumulated on the strip surface during hot rolling and storage. The scale is most commonly removed by *pickling* the strip while it passes through the tanks filled with hydrochloric or other types of acids.

Prior to descaling process, the strip is sometimes elongated by a tension *leveler* by several percent. The purpose is to facilitate the pickling by forming fine cracks in the scale. Skin pass rolling mill is also used in some pickle lines for the same purpose.

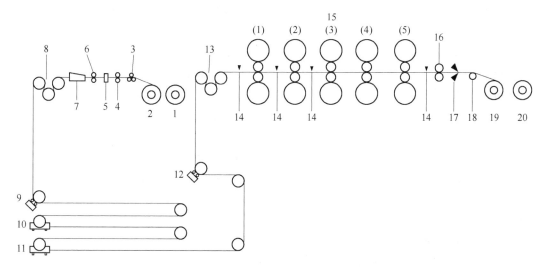

Fig. 4-9 Fully continuous cold mill

1—payoff reel No. 1; 2—payoff reel No. 2; 3—No. 2 flattener&pinch roll; 4—shear enter pinch roll; 5—shear; 6—shear exit pinch roll; 7—welding; 8—loop entry bridle; 9—No. 1 weld detector; 10—No. 1 loopcar; 11—No. 2 loopcar; 12—No. 2 weld detector; 13—mill entry bridle; 14—X-ray gages; 15—mill stands; 16—exit pinchroll; 17—parting shear; 18—detector rolls; 19—tension reel No. 1; 20—tension reel No. 2

(2) Cold rolling—High-production cold rolling is commonly conducted on tandem cold mills.

(3) Strip cleaning—During cleaning, the remaining of the rolling *solution* removed from the strip surface prior to *annealing*.

(4) Annealing and Cooling—Both batch type and continuous type annealing furnaces and cooling facilities are used to reduce *residual stresses* in the strip accumulated during cold rolling.

(5) Temper rolling—Single-stand non-reversing cold mills are customarily used as the temper mills.

(6) Finishing process—The finishing process involves final *leveling*, *slitting and cutting* the strip to the desired *delivery sizes*.

There is a strong trend in the steel industry toward integration of the processes described above. The most complete integration has been achieved by introducing a fully integrated processing line. This line incorporates all cold processing operations.

Words and Expressions

mill stand 轧钢机座
reduction 压下（量）
component 部件
work roll 工作辊
backup roll 支持辊
deflection 变形，压扁

roll gap adjustment mechanism 轧辊调整装置
pass line 轧制线
housing/mill housing 机架
main drive train 主传动机组
rolling torque 轧制力矩

roll arrangement 轧辊布置
two-high mill stand 两辊轧机机座
top, bottom and middle roll 上、下、中辊
cluster type mill stand 多辊轧机机座
intermediate roll 中间辊
asymmetrical 非对称的,非均匀的
symmetrical 对称的,均匀的
off-set roll 偏置辊
roll axes 轧辊轴向
horizontal mill stand 水平轧机机座
vertical mill stand 垂直轧机机座
crossed-roll mill stand 交叉轧辊机座
non-reversing mill stand 不可逆轧机机座
reversing mill stand 可逆轧机机座
back-pass mill stand 返回式轧机机座
direct drive 电机直接传动
gear drive 齿轮传动
gear box 齿轮箱
pinion stand drive 人字齿轮机座传动
independent drive 单独传动
planetary mill 行星轧机
rolling-drawing mill 轧-拔机组
reciprocating mills 往复式轧机
A_3 critical temperature A_3 临界温度
austenitic phase 奥氏体相区
warm rolling process 温轧过程
A_1 critical temperature A_1 临界温度
slabbing mill 板坯轧机
ingot 钢锭
plate mill 中厚板轧机
strip mill 带钢轧机
universal rolling mill 万能轧机
tandem roiling mill 连轧机
reheat furnace 再加热炉
roughing mill 粗轧机
transfer bar 中间坯
transfer table/delay table 中间辊道
finishing mill 精轧机
shear 剪切机
tandem finishing train 精轧连轧机组
looper 活套
interstand strip tension 机架间带钢张力
runout table 输出辊道
cooling header 冷却喷嘴
coiler 卷取机
descaling system 除鳞系统
scale breaker 破鳞机
roll coolant system 轧辊冷却系统
interstand cooling system 机架间冷却系统
semi-continuous 半连续
twin reversing roughing train 两机架可逆粗轧机组
three-quarter continuous 3/4 连续式
fully continuous 全连续
universal roughing mill stand 万能粗轧机座
single-pass stand 单道次机座
coilbox 卷取箱
Steckel mill 斯泰克尔轧机
hot direct charging rolling (HDR) (板坯)直接热装(炉)轧制
edge heater 边部加热器
heat insulator 热绝缘装置
reheating tunnel furnace 隧道式再加热炉
concast 连铸
thin continuous caster 薄板坯连铸机
equalizing furnace 均热炉
planishing mill 平整机
single-stand cold mill 单机座冷轧机
high-reduction 大压下
skin-pass 平整道次
temper mill 平整机
twin-stand/double-reduction cold mill 双道次压下冷轧机
tandem cold mill 冷连轧机
entry coil conveyor 入口带卷传送装置

entry coil lifting car 入口带卷升降小车
strip tension 带钢张力
tension reel 张力卷取机
provision 预留
pickling 酸洗
leveler 矫直机

solution 溶液
annealing 退火
residual stress 残余应力
leveling 矫直
slitting and cutting 纵切
delivery size 交货尺寸

5 Plate Mills

5.1 Introduction

Although one of the oldest types of hot mills, presently accounting for approximately 8% of steel shipment, plate mills have seen significant developments in recent years. They have undergone a series of evolutions, passing from the outmoded *universal mill*, to the *two-high, single-stand plate mill*, to the three-high mill and lastly to the modern *four-high configuration*. Modern single-stand plate mills now utilize larger housing, larger roll sizes and more drive power than any other type of mill stand. Typical of this trend is the plate mill installed at Dillingen, West Germany, which features rolls 1120mm and 2150mm by 4300mm driven by two motors rated at 8950kW each and capable of delivering a torque of 130 ton-meters.

These improved mills with their greater *rigidity* produce plates of more uniform thickness and superior flatness under both normal and controlled rolling conditions. In the later case, higher rolling forces and greater drive powers are required. Their performance is facilitated by modern instrumentation, automatic-gage-control (AGC) systems and computer control. Wedge-shaped plates (with the gage changing uniformly in the rolling direction) can now be produced for use in shipbuilding and for other purposes.

Finishing operations, such as the cutting of plates to size, their *inspection* and *marking*, are computer-controlled in the most recent plate-mill installations in Japan. Consideration is also being given to the in-line *quenching* and heat-treatment of plates following the last rolling pass as a means of conserving energy.

5.2 Types of Mills Used for the Rolling of Plates

During the last 200 years, various types of mills have been employed to roll plates. *Two-high pull-over mills* were used during the first half of the 19th century and *three-high and reversing mills* during the latter half of the same century. The universal mill, invented in 1848, was used soon thereafter in Pittsburgh, Pennsylvania, to roll plates 178mm to 610mm wide in thicknesses ranging from 5mm to 50mm.

As late as 1920, plates were being rolled on two-high reversing mills consisting of a train of two stands of plain rolls. In these mills the roughing stand nearer the steam engine was driven through both rolls whereas the finishing stand was driven only through the bottom roll. In many cases, especially in England, the rolls were run hot, no attempt being made to cool them. *Two-high non-reversing mills* are now obsolete for producing finished plate but two-high mills are still used as scale-breakers and roughers, *in tandem with* a finishing unit which is either a three-high or a four-high mill.

A wide variety of tandem plate-mill layouts exist because such arrangements often represent modifications to original rather than new installations. Such facilities may be categorized as semi-continuous and continuous, the former utilizing multi-pass reversing roughing units whereas the latter use non-reversing roughing units coupled with two or more *single-pass* stands in which the plate is reduced simultaneously. Two-, three- and four-high stands with or without *scale breakers*, and *edgers* are used as roughing stands while four-high mills are used as finishing units.

The use of a light vertical stand for edging purposes was widespread during the 1950 to 1965 period. However, due to economic reasons, this practice is not incorporated in the newer mills, but plate mills built for the rolling of heavy slabs still utilize edger.

5.3 Plate-mill Design

Whereas early plate mills often rolled small slabs into plates, modern plate mills generally utilize reheated, *conditioned slabs*. The latter approach permits the attainment of better surface quality in the rolled product, better *mechanical properties* in the plate through higher reduction ratios between initial and final thicknesses, improved temperature control in the rolling operation and greater output because of fewer passes. Many plate mills are now rolling, or will be rolling in the near future, continuously cast slabs, where a reduction ratio of four or five to one is required. Thus, when continuous cast slabs up to 250mm to 300mm thick are available, the entire thickness range of plates (up to 76mm) can be obtained. However, to ensure maximum mill productivity, the slab thickness should be directly proportional to plate thickness. For this reason, continuous casters may have to be capable of convenient adjustable of slab thickness as, for example, by the use of in-line *reduction stands*.

To reheat the conditioned slabs, the most modern conventional or *walking beam furnaces* are now used. Such furnaces are not only thermally efficient, but provide a high degree of temperature uniformity in the heated slabs. They should have a maximum capacity about 50% greater than the average output of the plate mill.

Modern plate mills are capable of rolling product at least 2540mm wide and often up to 5080mm in width. In some cases, plates are rolled double-width and later cut to *ordered width*. Typically, both roughing and finishing stands are utilized. *Scale breaking stands* are often incorporated in the older plate mills, mills built after 1966 have no special *horizontal or vertical decaling stands*. This has resulted from the successful use of high-pressure water sprays to remove both *primary and secondary scale*. Sometimes, to maintain constant spacing from top surface of the workpiece, the spray headers are attached to the *top work- or top backup-roll chock*. On the roughing stands, they are often located on each side of the stand and, on the finishing stands, they are placed only on the entry side. In order to avoid excessive plate cooling, some mills utilize steam descaling.

An increasing traction of plate output is now controlled-rolled, and some two-stand mills feature a cooling bed connected to the *mill table*. In such cases, the workpiece may be moved backwards and forwards to prevent localized cooling by the *table rolls*.

To achieve the desired stiffness, modern mills use backup rolls of large diameter (often in excess of 2030mm) and *housing posts* of large cross-sectional area (often in excess of $0.97m^2$). The weights of the mill housings often exceed 250 tons and mill *modul* of the order of $1×10^6 t/m$ are attainable. *Specific rolling forces* in the range of 1070t/m to 2140t/m are generally achievable and thinner plates (down to about 28mm in thickness) may be conveniently rolled.

Work-roll diameters are established by the maximum torque required to transmit the deformation energy to the workpiece. Mills over 4060mm in width are designed for torque of over 576t/m and use work rolls 1020mm to 1120mm in diameter. Generally, the ratio of torque to roll-body length is 113400kg · m/m to 158760kg · m/m.

Many of the new mills incorporate two *roll-positioning systems*. A *high-speed screw down system* (maximum speed 910mm/min to 1830mm/min) is utilized for adjusting the *roll gap* between passes and a low-speed, high-accuracy system (maximum speed 76mm/min to 152mm/min) is used for adjusting the rolls to a positional accuracy of one *mil* under load for the purpose of automatic gage control. Where a mill may be used either as a *blooming mill* or a plate mill, separate drives may be utilized with the screwdown system. The latest plate mills incorporate *hydraulic roll positioning mechanisms* and, for the control of *plate profile and flatness*, *work-roll and/or backup-roll bending systems* are utilized.

Computers are now generally used for the control of new plate mills, as exemplified by the two-stand 32mm mill at the Hoesch Dortmund Works. The computer control of this mill ensures (1) thickness uniformity and flatness of the plate, (2) a high degree of width uniformity and decreased *edge scrap* through *width control* of the roughing stand, (3) increased output through *material tracking*, *process control* and the optimum balance of the passes between the two stands and (4) a decreased number of rolling passes. To be able to perform these functions, the computer receives signals from 41 *photocells* mounted along the roller tables (indicating the position of each workpiece) and non-contacting measuring devices for monitoring workpiece temperature, thickness, width and length, as well as the rolling force and spindle torque.

An important consideration in the operation of plate mills is the handling of scale removed from the slab furnaces. In some cases, the scale-removal system is built under the tables and mill stands as part of the foundation.

Words and Expressions

universal mill　万能轧机
two-high, single-stand plate mill　两辊单机架中厚板轧机
four-high configuration　四辊配置（轧机）

rigidity　刚度
inspection　表面检查
marking　打印
quenching　淬火
two-high pull-over mills
　　两辊往复式轧机
three-high and reversing mills
　　三辊轧机和可逆轧机
two-high nonreversing mill
　　两辊不可逆轧机
in tandem with　与……连轧
multipass　多道次
single-pass　单道次
edgers　立辊轧机
conditioned slab　精整板坯
mechanical properties　机械性能
reduction stand　压下机架
walking beam furnaces　步进梁式加热炉
ordered width　定货宽度
scale breaking stand　除鳞机架
horizontal or vertical descaling stand
　　水平和垂直除鳞机架
primary and secondary scale
　　初生和二次氧化铁皮

top work-or top backup-roll chock
　　上工作辊或上支持辊轴承座
mill table　轧机辊道
table roll　辊道辊子
housing post　机架立柱（牌坊）
modulus　系数，模数
specific rolling force　单位轧制压力
roll-positioning system　轧辊定位系统
high-speed screw down system
　　高速压下系统
roll gap　辊缝
mil　密耳（1mil = 10^{-3} inch）
blooming mill　方坯初轧机
hydraulic roll positioning mechanism
　　液压轧辊定位装置
plate profile and flatness
　　中厚板断面和板形
work-roll and backup-roll bending system
　　工作辊和支持辊弯辊系统
edge scrap　切边
width control　宽度控制
material tracking　轧件跟踪
process control　过程控制
photocell　光电管

6 Hot-Strip Mills

6.1 Introduction

No type of mill has probably experience more dramatic developments than the hot-strip mill during the 1950's and 1960's. Following the building of "*Generation* I " mills in the 1930's and the 1940's, the succeeding decades saw the introduction of "*Generation* II " mills of greater width and increased drive power, with *tandem roughing stands*, *7-stand finishing trains*, improved *automatic-gage-control systems*, *laminar-flow cooling* on the *runout tables*, the use of *hot-rolling lubricants* and the practice of "*zooming*" or the gradual increase in rolling speed during the processing of coil. The newer mills featured complete computer control of the *reheat furnaces*, the *roughing* and *finishing stands* and the cooling sprays on the runout table. Although capable of rolling strip down to a thickness close to 1mm, economic reasons usually *dictate* minimum finish gages of about 2mm or thicker. At the higher end of the hot-strip mills for *spiral-pipe* manufacture and other applications.

Still further innovations have characterized the few new "*Generation* III " mills installed during the 1970's. These have included improved instrumentation (including *shape and flatness sensing*), the computer control of the finishing train to provide the desired cross-sectional profile and flatness of the rolled strip and the *Stelco coilbox* installed between the last rougher and the finishing train (which coils the bar after roughing and minimizes the length requirement of the *delay table*) .

Unfortunately, however, the total costs of Generation II and Generation III hot-strip mills have been so large as to limit the number of installations in recent years and to encourage not only innovations with respect to conventional hot-strip mills but also a *resurgence* in the use of simpler, more economical mills, particularly for the rolling of *specialty steels*. The latter includes the hot *Steckel mill* and relatively new types of mills, such as *planetary mills*.

6.2 Steckel Hot Mills

A sketch illustrating the general arrangement of a Steckel hot mill is shown in Fig. 6-1. It consists of a single 4-high reversing mill stand with two coilers located above the *passline* and contained within furnaces. In some cases, a slab 76mm to 127mm thick is rolled back and forth in the mill until a thickness of about 13mm has been obtained. Then the workpiece is directed up to a furnace coiler and the temperature of the coil is maintained at about 898°C for the successive passes. In other installations, a *two-high roughing stand* is used ahead of the Steckel mill to convert the small slabs to bars. After roughing, the bar passes through a *tunnel-type furnace* into the Steckel mill and then to the coiler on the remote side of the mill. In the successive passes, the bar is reduced to *hot-strip gage* and is then transferred out to a conventional *cooling table* and coiler. Fig. 6-1 shows the layout of a recent Steckel mill installation used primarily for rolling stainless steels. Note here the use of an induction furnace for heating the slabs.

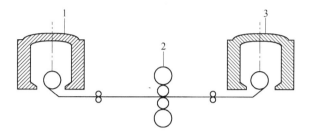

Fig. 6-1 Schematic diagram of Steckel mill
1—left furnace coiler; 2—four-high mill; 3—right furnace coiler

Roll wear is more severe with this type of hot-strip mill than with semi-continuous and continuous hot-strip mills. However, this disadvantage can be *alleviated* considerably by the use of *hot-rolling lubricants*.

Steckel hot mills are capable of productivities ranging from 250000 tons to over 1000000 tons per year. Only a limited number are now in operation throughout the world, since some of the earlier mills have been converted into semi-continuous facilities. They are principally used by small, independent producers to hot roll relatively small tonnages of stainless steel or other high-value slabs to strip.

6.3 Planetary Mills

The basic principle of operation of a planetary mill is illustrated in Fig. 6-2 and a typical mill layout consist of feed rolls, the planetary roll clusters, a looping system and planishing rolls. The mill is capable of providing up to a 20∶1 *reduction ratio*. However, for economic and metallurgical reasons, hot-strip mills should be capable of providing at least a 75 : 1 overall reduction ratio. Hence planetary mills must be supplemented with roughing and/or finishing stands.

Because of the slow entry speed of the workpiece into a planetary mill, it is usually desirable to locate a *roller-hearth furnace* close to the mill. When the mill is used in conjunction with a continuous caster, the slabs are already hot and only require *soaking*.

The action of the small planetary rolls is to produce a "washboard" type of surface on the strip. This surface characteristic must be removed by the use of a two-high or four-high planishing mill following the planetary mill. Where the latter type of planishing mill is used, an additional 30% reduction may be given to the strip.

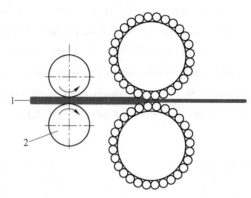

Fig. 6-2　Layout of planetary mill
1—slab; 2—feed roll

6.4　Semi-continuous Hot-strip Mills

Such installations generally consist of two or more *slab reheat furnaces*, *a two-high scale breaker* (*and/or a hydraulic descaling unit*), a powerful *reversing roughing stand*, a *finishing train with four or more 4-high stands*, a long *runout table* and one or more coilers. This arrangement is primarily intended for production rates of about 1.5 million tons to 2 million tons per year and where the slabs to be rolled have a comparatively low weight per unit width. It is a very flexible arrangement and can permit a varying number of roughing passes. Accordingly, it can accept slabs of differing *steel grades* and thicknesses which require rolling with different numbers of roughing passes.

　This type of mill, however, is not particularly suitable for rolling slabs of large weight per unit width as the length of the mill required for such workpieces becomes very extended and the capital cost per annual ton of output becomes very high. When rolling low specific weight slabs, the total roughing time in the reversing rougher usually exceeds the rolling time in the finishers, making the reversing rougher the limiting factor in output calculations.

　Sometimes, this type of mill in Fig. 6-3 is built with future expansion in mind as, for example, the addition of more finishing stands.

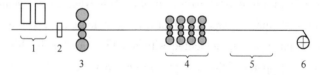

Fig. 6-3　Layout of semi-continuous hot-strip mill
1—slab reheat furnace; 2—edger; 3—roughing mill; 4—finishing mill; 5—runout table; 6—coilers

As a typical example of a semi-continuous mill, Wheeling-Pittsburgh Steel Corporation's 1680mm hot-strip mill at Allenport, Pennsylvania, may be cited. It consists of: (1) a 2-high scalebreaker with 910mm by 1680mm rolls driven through a *pinion stand*, *gear reducer* and *flywheel*, (2) a first descaling header operating at 8270kPa on the discharge side of the scalebreaker, (3) a *four-section table* between the scalebreaker and the rougher with each section independently driven, (4) a 4-high reversing rougher using rolls 910mm and 1240mm by 1680mm, preceded by a vertical edger with 840mm diameter rolls, (5) a three-section table between the rougher and the *crop shear*, (6) a crop shear of the drum type, (7) a finish scalebreaker which includes pinch rolls, a table and two descaling headers, (8) a finishing train of four 4-high stands using 640mm and 1240mm by 1680mm rolls with a delivery speed of 540m/min, (9) a three-section runout table 113m long, each section being equipped with banks of sprays and 59 independently driven rollers, (10) a *downcoiler* featuring a 910mm diameter mandrel, four pairs of motor-driven *wrapper rolls* and a set of pinch rolls (the top roll being 910mm in diameter and the bottom roll 410mm in diameter), (11) a *downender* which deposits the coils on end on a transfer conveyor, (12) *pyrometers* which indicate and record the slab temperatures and the temperatures of the bars on the delivery side of the scalebreaker, ahead of the crop shear, between the last two finishing stands and at the downcoiler, (13) two *oil cellars*, one in the roughing housing pumps for the descaling sprays, the *hydraulic roll-balance systems* and the *lubrication systems* for the mill drives and *roll bearings*, (14) seven *automatic grease systems* and (15) two reheat furnaces each rated at 80 tons per hour.

6.5 Continuous Hot-strip Mills

In these mills, the slab is processed only in one direction through the mill, there being no reversals at any stands. Initially, the roughing stands were spaced at sufficient distances so that the slab would be rolled in one stand only at any given time. More recent installations, however, utilize close-coupled roughers which operate in tandem.

The general layout of such mills is inflexible because the number of roughing passes is fixed by the number of stands and little provision can be made in initial layouts for future expansion. Initial capital costs are high because of the number of mill stands and the overall length of the mill layout. However, the rolling time in the last rougher is always less than the rolling time in the finishing train. Consequently, the very highest outputs (close to 5 million tons per year) are possible with continuous hot-strip mills.

In the normal roughing train layout, the first one or two mills are 2-high and the remainder 4-high stands. Roughing stands are usually driven by *synchronous motors* because of their good speed-torque characteristics. Exit speed from the last rougher may be as high as 4 meters per second or higher than the normal entry speed into the finishing train. Cooling water is generally applied to both the work and backup rolls on the exit side of each stand. Except for the first Stand, all roughers usually have vertical edging rolls on their entry side, the diameters of these being about 915mm. These rolls are capable of taking edge reductions up to about 25mm on the slabs.

Ahead of the finishing train is a crop shear followed by a 2-high scalebreaker and high-pressure descaling sprays. Before these units, however, some mills have been equipped with *radiation shields* or, in other cases, tunnel-type furnaces to conserve the thermal energy in the bar as much as possible.

With respect to finishing trains, all the mill stands are four-high and are generally of identical design spaced at approximately 5.5 meter centers. Work rolls are about 710mm and backup rolls about 1520mm to 1580mm in diameter. Each stand is driven with motors with power in the range 7460kW to 8950kW, except for the last stand which is usually driven by about 5970kW. Mill-stand speeds are usually controlled by *interstand loopers* with one finishing stand, usually the fourth, being the *pivot or reference stand*. Adequate roll cooling must be employed to maintain good roll life and to control both the *crowns* of the rolls and the cross-sectional profile of the hot-rolled strip. The *screwdown systems*, whether *hydraulic or electromechanical*, must be capable of operation *under load* so that automatic-gage-control systems may be effective.

As an example of a continuous hot-strip mill constructed in the 1960's, the 2130mm mill at the Gary Works of U.S. Steel Corporation may be given. The mill, started up in 1967, utilizes four *continuous 5-zone slab-heating furnaces* each with a capacity of 336t/h. Slab sizes rolled range in width from 460mm to 1940mm, in thickness from 100mm to 250mm, with a maximum length of 12200mm and a maximum weight of 47 tons. The thickness of product rolled lies in the range 1mm to 13mm over the complete range of slab sizes.

The layout of the mill is shown in Fig. 6-4. It is to be noted that the fourth and fifth roughers are close-coupled with the fourth rougher being driven with a D.C. motor. The delay table is 126mm long and features 139 rollers 356mm in diameter individually driven by 2kW, D.C. motors. The runout table is 152m long with 336 rollers 305mm in diameter on 460mm centers individually driven by 4kW, D.C. motors at speeds up to 1220m/min. Since installation, the runout table has been equipped with a *laminar-spray cooling system*.

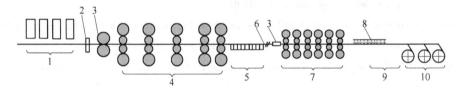

Fig. 6-4 Layout of a continuous hot-strip mill
1—furnace; 2—VE; 3—scale breaker; 4—roughing mill; 5—transfer table;
6—shear; 7—finishing mill; 8—water coolant headers; 9—run-out table; 10—coilers

Three *mandrel-type downcoilers* used in the installation are capable of forming coils with an internal diameter of 720mm and an outside diameter up to 2030mm.

For descaling purposes, *centrifugal pumps*, each driven by a 2240kW motor, supply high-pressure water to sprays located at the vertical scalebreaker, the entry sides of the No. 3, No. 4 and No. 5 roughers, the finisher descaling unit and the exit sides of finishers No. 1, No. 2 and No. 3.

A number of mills of comparable capacity were constructed in the last two decades, each being capable of rolling 3 million tons to 5 million tons per year. Such mills are generally referred to as being of the "Generation II" or the "Generation III" type. However, because of the enormous

6.6　Rougher Trains

Because of the necessity of eliminating scale from the surfaces of slabs prior to conversion into strip, roughing trains generally feature a scalebreaker as the first mill stand. This unit deforms the slab sufficiently to break and loosen the scale, which is then blasted from the slab surface by high-pressure jets of water or steam. Whereas the early scalebreakers were generally of the horizontal type, all of the latest hot-strip mills use roughing scalebreakers of the independent, vertical edger type capable of reducing the width of thick slabs by 50mm to 100mm. Rolls are generally of *alloyed cast steel* about 1140mm in diameter. The *cast-steel housings* are symmetrically positioned with respect to the center of the passline. Occasionally, vertical scalebreakers are found in combination with *so-called* horizontal scalebreakers but, as the latter are similar in design to the roughers, they are commonly considered to be roughing stands.

Because some of the early hot-strip mills were appreciably wider than the slabbing mills then installed, roughing trains of hot-strip mills often incorporated a *broadside-mill stand*. This unit enabled the width of the slab to be increased by rolling the slab crosswise. Such mills were of the 2-high and 4-high types and were equipped with *turntables*, usually on both sides of the stand, for turning the slabs so that they could enter the mill stand edgewise. To ensure that the slabs entered the mill properly after being turned broadside to it, the mills were also generally equipped with *slab pushers*.

Roughing stands are of the 2-high and 4-high types and, until recently, were placed sufficiently far apart that a workpiece could only be rolled by one rougher at a time. The 2-high mills, with rolls about 1270mm in diameter, offer a *bite angle* permitting very *heavy drafts* in the early roughing passes. The semi-continuous and the three-quarter type mills use a reversing rougher whereas in continuous mills the roughers (up to six in number) are nonreversible and are powered by synchronous motors. Such drives range from 3730kW to 4470kW for the first rougher to 7460kW to 11930kW for each of the last two stands.

With the exception of the first, and sometimes the second, all roughing stands have attached vertical edgers on their entry sides. The edgers use *cast alloy steel rolls* 840mm to 970mm. The D. C. drive motors range in power from 112kW to 373kW. The roll surface speed of the edgers must be adjusted to correspond to that of the horizontal rolls and hence operate under adjustable voltage control.

In some instances, the last two roughers are close-coupled, the *penultimate stand* being driven by a D. C. motor, the speed of which is appropriately matched to that of the last rougher so as to provide a "tension-free" condition to the bar as it is simultaneously reduced by he two stands. The term "tension-free" implies a tensile stress of not more than about 0.5 kg/mm^2 and may be achieved by the use of a interstand looper.

6.7 Finishing Trains

Finishing trains of modern hot strip mills consist of six or seven closely coupled (about 5490mm between centers) four-high mill stands of essentially identical construction. *Housing posts* are about 0.71m² in cross section, workroll diameters are usually 720mm to 760mm and backup rolls usually feature diameters in the range 1520mm to 1570mm. The work rolls are supported by *roller bearings* (sometimes *mist-lubricated*) whereas the backup rolls are contained in *oil-film bearings*. Top backup rolls may be balanced by *hydraulic cylinders* in the *chocks*. Similarly, the top work rolls may be balanced by hydraulic cylinders in the bottom *work-roll chocks*. *Work-roll bending systems*, capable of applying forces up to 25 tons on each *roll neck* are often installed in the finishers and, in one installation, *backup roll bending* is provided at the last stand.

In all recent hot-strip-mill installations in the U.S.A., conventional screwdown systems have been used with *screws* of 460mm to 530mm in diameter, with *pitches* ranging from 19mm to 38mm, driven by two D.C. motors of 37kW to 74kW under adjustable voltage control. In the finishing stands, screw speeds of 38mm/min to 76mm/min are typical. However, *hydraulic roll-positioning systems* have been used on some mills.

Entry guides are used at each stand. Loopers (some of which are hydraulic) and tables are provided between each pair of stands and an *exhaust system* may be used with some of the last stands to remove oxide dust. At the end of the train, there is usually a *strip-thickness and a width gage*. Roll-cooling spray headers are used at all stands and, in recent years, rolling-lubricant application systems have been installed.

Automatic roll-changing equipment is now generally installed on the newer hot-strip mills, this equipment enabling work rolls to be changed in 10min.

Words and Expressions

Generation I　第一代带钢热轧机
Generation II　第二代带钢热轧机
tandem roughing stands　粗轧连轧机组
7-stand finishing train　七机架精轧机组
automatic-gage-control system
　厚度自动控制系统
laminar-flow cooling　层流冷却
runout table　输出辊道
hot-rolling lubricant　热轧润滑
zooming　加速轧制
reheat furnace　再加热炉
roughing stand　粗轧机座

finishing stand　精轧机座
dictate　确定
spiral-pipe　螺旋管
profile　（带钢）横断面
shape and flatness sensing
　板形和平直度传感器
Stelco coilbox　Stelco 卷取箱
delay table　中间辊道
resurgence　苏醒
specialty steel　特殊钢
Steckel mill　Steckel 轧机
planetary mill　行星轧机

passline 轧制线
tunnel-type furnace 隧道式加热炉
two-high roughing stand 两辊粗轧机座
hot-strip gage 热带厚度
cooling table 冷却辊道
roll wear 轧辊磨损
hot-rolling lubricant 热轧润滑
alleviate 缓解，减轻
reduction ratio 压下比
roller-hearth furnace 辊底式炉
soaking 均热
slab reheat furnaces 板坯再加热炉
two-high scale breaker 两辊破鳞机
hydraulic descaling unit 液压除鳞装置
reversing roughing stand 可逆粗轧机座
finishing train with four or more 4-high stands
　由四个或四个以上四辊轧机组成的精轧机组
runout table 输出辊道
steel grade 钢种
pinion stand 人字齿轮机座
gear reducer 齿轮减速箱
flywheel 飞轮
four-section table 四段式辊道
crop shear 切头飞剪
downcoiler 地下式卷取机
wrapper roll 助卷辊
downender 翻卷机
pyrometer 高温计
oil cellar 油站
hydraulic roll-balance system 液压轧辊平衡系统
lubrication system 润滑系统
roll bearing 轧辊轴承
automatic grease system 自动油脂系统
synchronous motor 同步电机
radiation shield 辐射罩
interstand looper 机架间活套
pivot or reference stand 基准机架
crown 凸度
screwdown systems 压下系统

under load 带负荷
hydraulic or electromechanical
　液压或机械电动（压下）
continuous 5-zone slab-heating furnace
　五段连续式板坯再加热炉
laminar-spray cooling system 层流冷却系统
mandrel-type downcoiler
　芯轴式地下卷取机
centrifugal pump 离心泵
roller bearing 滚动轴承
mist-lubricated 油雾润滑
oil-film bearing 油膜轴承
hydraulic cylinder 液压缸
chock 轴承座
work-roll chock 工作辊轴承座
roll neck 辊颈
work-roll bending system 工作辊弯曲系统
backup roll bending 支持辊弯曲
alloyed cast steel 合金铸钢
cast-steel housing 铸钢机架
so-called 所谓的
broadside-mill stand 宽展机架
turntable 转向辊道
slab pusher 板坯推钢机
bite angle 咬入角
heavy draft 大压下
cast alloy steel roll 铸钢轧辊
penultimate stand 倒数第二个机架
screw 压下螺丝
pitch 螺距
hydraulic roll-positioning system
　液压轧辊位置系统
entry guide 入口导板
exhaust system 除尘系统
strip-thickness and a width gage
　带钢厚度和宽度仪表
roll-cooling spray header 轧辊冷却喷嘴
automatic roll-changing equipment
　自动换辊装置

7 Twin-roll Casting Technology

7.1 Introduction

BHP (Broken Hill Proprietary Company) of Australia and IHI (Ishikawajima-Harima Heavy Industries) of Japan began a collaborative development project in 1988 to commercialize *twin-roll*, *strip casting* of steel. Originally patented by Sir Henry Bessemer in 1857, the concepts for twin-roll casting of metal products are not new, as metallurgists have long sought methods for casting metals at or *near their final shape*. Code-named Project "M", BHP and IHI built a *pilot plant* in Port Kembla, Australia with the capability of casting 5 - tonne coils 800mm in width. Initial results with *austenitic stainless steels* were *promising*. By 1991, commercially acceptable *304 stainless coils* were cast at 800mm widths and subsequently rolled at BHP Steel's stainless steel facilities. Initial attempts at casting *carbon steel* indicated that this would be much more difficult. The pilot plant successfully cast a series of 1300mm wide, *low-carbon heats* in 1992.

In 1993, the Board of Directors of BHP approved the construction of a full-size "Development Plant" aimed at proving the technical viability of strip casting of carbon steels. Construction began in 1993 and the first casts were conducted in early 1995. This plant was "full size" in that it was capable of casting 60-tonne *ladles* of carbon steels into maximum 30-tonne coils, with a width of 1345mm (although the equipment was capable of widths to 2000mm wide). Initially, the target thickness was 2.5mm. Heat sizes for the plant were limited by the *order*, low-power *EAF* (Electric Arc Furnace) utilized for the project.

The Project "M" Development Plant was operated until the end of 1999, perfecting the technology of casting low-carbon steels into a variety of commercial products. In total, more than 35000 tonnes were produced at the plant, *culminating with* a process capability trial of 29 heats at 50 tonnes each, at a cast thickness between 1.9mm and 2.0mm. The material was subsequently *cold-rolled*, *metallic-coated*, and painted into commercial *roofing products* for the Australian market. Additional material from the run was shipped to a BHP *subsidiary* in the United States where it was processed into structural decking for the construction market. *CASTRIP* material has also been processed into *pipe and tube* of various diameters. These results clearly proved the technical feasibility of the CASTRIP process. However, without a supply of *molten steel* to conduct *multi-heat* sequences, full commercial feasibility could not be determined.

During 1999, a Commercialization Team was formed within BHP to carefully examine the most appropriate path forward for commercializing the BHP/IHI technology. The status and future of the industry was examined and the alternatives considered ranged from simply ending the technology development, through construction of a BHP-owned, full-size commercial plant. The final recommendation was to invite the participation of a third partner to bring the CASTRIP technology to full-scale commercial production.

A worldwide search for the ideal partner was conducted and Nucor Corporation, the largest steel producer in the United States, was identified as the optimum choice. Nucor possesses an excellent track record in technology implementation combined with significant experience in flat-rolled steel production. Discussions among BHP, IHI and Nucor led to the formation of *Castrip LLC*, which is now established in Charlotte, North Carolina (Nucor's headquarters). Castrip LLC is a *limited liability company* owned 47.5% by Nucor, 47.5% by BHP and 5% by IHI. The purpose of Castrip LLC is to make the technology and patents related to the CASTRIP process available to third parties. Nucor is the first *licensee* of the CASTRIP technology through Castrip LLC and is now building the first truly commercial strip casting plant in the world at Crawfordsville, Indiana. When the Nucor CASTRIP plant is operating, cumulatively more than US $ 400 million will have been spent on the development from laboratory through commercialization.

7.2 Process Overview

The CASTRIP process is based upon the same concepts that Sir Henry Bessemer patented in the mid-19th century. Fig. 7-1 shows a simple schematic of the basics of the process—two counter-rotating rolls that provide a surface or *mold* against which molten steel solidifies. As the Fig. indicates, the steel begins to solidify against the rolls just below the *meniscus* and shell growth continues as it moves downwards through the melt pool. At the *roll nip* or *pinch point*, the two shells are essentially fused together forming a continuous strip which then exits the caster in a downward direction.

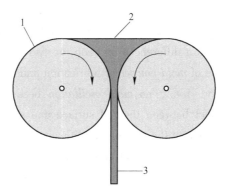

Fig. 7-1 Simple schematic of twin-roll strip casting process
1—couter-rotating rolls; 2—liquid metal; 3—solid strip

Although the concept is relatively simple, its application at a commercially production level has proven to be extremely difficult. Several technical advancements have occurred in recent years that

have made twin-roll casting possible at a commercial level. These include:
(1) High speed computing and process control;
(2) Advanced *ceramics* and materials (including copper alloys);
(3) Sensing technology;
(4) *Mathematical modeling* of physical phenomena.

In addition to these advancements, the Project "M" team had to significantly increase the body of knowledge related to several key areas of *process metallurgy* directly connected to the twin-roll process. Previous papers have discussed the development of the Castrip process; these advancements or *breakthrough* are what sets CASTRIP technology apart from other twin-roll processes and can be divided into 5 key areas:
(1) *Metal delivery*;
(2) Early solidification;
(3) *Edge containment*;
(4) *Roll distortion*;
(5) *Refractories*.

Metal delivery. Metal delivery to the melt pool is critical for a number of reasons. Unlike *conventional casting*, the melt pool is very small in the CASTRIP process. As a result, the ratio of mass flow rate into the pool to the pool volume is nearly of order of magnitude higher in the CASTRIP process compared with *slab casting*. Thus, the metal delivery nozzle utilized in the CASTRIP process is completely different from that used for *conventional casting*, with major emphasis on reducing the turbulence in the steel as it enters the melt pool and the need for effective distribution of metal along the roll length. Any disturbance at the *meniscus* invariably manifests itself as a strip defect. Thus the process must be stable and in control at all times to ensure excellent surface quality.

Early solidification. Most of the effort related to the understanding of solidification in steels has been confined to continuous casting over the past 20 years. There are some major differences between CASTRIP technology and slab casting that have significant effects on the formation and growth of the *shell*. Among the main differences between the two processes are that the CASTRIP technology does not use any type of *mold powder* or lubrication and that the mold (roll) and steel shell maintain the same velocity, i.e., no *mold oscillation* because there is no mold powder, there is significantly better contact between the roll surface and solidifying shell, starting at the meniscus and extending down to exit at the roll nip. Significant work has been done in trying to understand the mechanisms for shell formation and growth as well as *heat transfer* between the steel and roll surface. This work has been described previously and it is critical that the variables affecting the early solidification of the shell and its subsequent growth be understood for the production of quality strip.

Edge containment. Although most of the surface area of the solidifying strip is confined to contact against the face of the rolls, the edge containment of the *melt pool* proved to be a technical challenge that required significant focus during the development of the CASTRIP process. The problem is related to the fact that freezing is most likely to occur in this area because of heat loss through the *side dam material* as well as through the rolls. *Premature freezing* can lead to poor edge quality as well as *triggering* a series of events that eventually lead to the cessation of casting. Many materials were tested for use as side dams before a suitable refractory was found.

Roll distortion. Roll distortion is caused by the generation of *thermal stresses* as the casting rolls become heated by the solidifying steel. The proper design of casting rolls must take into account this unavoidable distortion in order to produce strip with a desired thickness profile. Previous publications have demonstrated the variation in *heat flux* and temperature across a casting roll. This work also showed that rolls with a machined in *crown*, where the center is smaller in diameter than the edges and will provide an optimum *strip profile*.

Refractories. The interaction between the liquid steel in the melt pool and the refractories utilized for the metal delivery nozzle has been shown to cause defects in the strip cast material. *Active oxygen* contained in the pool was shown to combine with carbon in the *alumina graphite nozzle*, resulting in the formation of *CO bubbles*. These bubbles caused disturbances at the steel surface and at the meniscus where they resulted in surface defects on the strip. Defects caused by this mechanism have been eliminated through the selection of appropriate refractory materials.

7.3 Nucor's Crawfordsville CASTRIP Facility

Construction of the Nucor CASTRIP facility began in February 2001, with the *start-up* of the plant expected in May 2002. Fed from an existing EAF shop located less than 1km away, the CASTRIP plant includes a *ladle metallurgy furnace* (*LMF*) to make the necessary temperature and chemistry adjustments prior to casting. Fig. 7-2 shows a sketch of the general layout of the Nucor CASTRIP caster building. In overall dimensions, the building is approximately 135m wide and

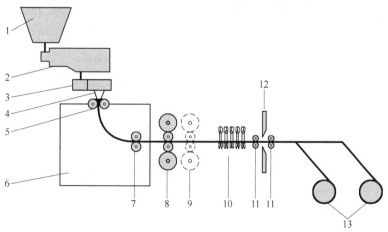

Fig. 7-2 A sketch of the general layout of the Nucor CASTRIP caster building

1—ladle; 2—tundish; 3—transition piece; 4—delivery nozzle; 5—casting rolls; 6—controlled atmosphere; 7—pinch rolls; 8—hot rolling stand; 9—2nd hot rolling stand (optional); 10—cooling table; 11—shear; 12—pinch rolls; 13—coilers

155m long. The total length of the casting operation, from *ladle turret* to coilers is only 60m. This is in contrast to a *slab caster* with a *reheat furnace* and *hot strip mill* that typically requires 500m to 800m of length to make the same hot-rolled products.

The CASTRIP facility is shown in profile in Fig. 7-3, indicating the main features of the process. The ladle and *tundish* are based on standard steel industry designs. The caster does not use a *dummy bar* for start-up. Upon exit from the casting rolls, the solidified strip is directed to a *pinch roll* then through a *hot rolling stand*. In this transition, the atmosphere is controlled to limit oxidation of the strip and the formation of *scale*. Equipment from the hot rolling mill through the coilers is also of standard steel industry design.

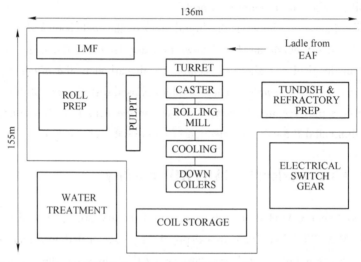

Fig. 7-3 Nucor CASTRIP plant layout

General specifications for the Nucor CASTRIP caster are shown in Table 7-1. Of particular note is the diameter of the casting rolls—500mm. This diameter is significantly smaller than in other twin-roll casting projects and one of the main advantages of the CASTRIP process.

Table 7-1 Nucor CASTRIP plant specifications

Unit	Specification (in metric units)
building dimensions/m	155×135
heat/ladle size/t	110
caster type	500mm diameter twin roll
casting speed/m · min^{-1}	80 (typical), 150 (maximum)
product thickness/mm	0.7~2.0
product width/mm	2000 maximum
coil size/t	25 (typical), 40 (maximum)
in-line mill	Single stand-4 High with Hydraulic AGC
work roll dimensions/mm	475×2050
back-up roll dimensions/mm	1550×2050

Continued 7-1

Unit	Specification (in metric units)
rolling force/t	3000 (maximum)
main drive/kW	3500
cooling table	10 top and bottom headers
coiler size/t	2×40 coilers
coiler mandrel/mm	760 diameter
annual capacity/t · a^{-1}	500000

Smaller roils are less expensive to build, have a lower operating cost and are capable of producing thin-quality cast products. The Nucor CASTRIP plant will have the capability of casting steel strips at 1.0mm to 2.0mm in thickness, plus a single hot-rolling stand to reduce the material by another 30%. At casting speeds of about 80m/min, average cast thickness of 1.6mm, and average strip width of 1211mm, the output of the plant is expected to be about 500000 tonnes per year.

Although the initial casting of UCS steel at the Nucor CASTRIP plant will be low-carbon steels—and this is by far of the greatest commercial interest worldwide—Nucor also plans to begin casting stainless steel within the first year of operation. Development work recently undertaken indicates that 409-grade stainless steel should be well within the casting parameters of the Nucor CASTRIP plant. Combined with BHP's earlier development work, it is expected that a single CASTRIP plant will be able to easily cast carbon, austenitic and ferritic stainless steels. In general, the development path has indicated that a twin-roll casting machine designed and built for carbon steel will be able to cast stainless steels, but the reverse is not true.

7.4 The Limits of Present Thick and Thin Slab Casting

In the fundamental study work done by the BHP Commercialization Team in 1999, the following situation analysis for the adoption of the CASTRIP process was established:

(1) The world steel flat-rolled industry is characterized by cumulative overcapacity with regional undercapacity.

(2) For the production of conventional hot-rolled, low-carbon steels (>2.5mm thickness), both thick and thin slab casting is a rational production train with potentially attractive financial returns in "normal" circumstances.

(3) For the production of Ultra-Thin Gauge Hot-Rolled (UTGHR<2.0mm thickness) low-Carbon steels, both thick slab and thin slab casters followed by tandem hot mills are technically feasible production methods, but both suffer from unattractive production costs.

(4) For end-product thicknesses of less than about 1.3mm, the production costs of either thick or thin slab casters escalate to the point that pickling and cold rolling is, and almost certainly will continue to be the most practical production method.

(5) Long-term, overall flat steel consumption will continue to rise at an average of 2% per year, but demand for UTGHR will rise at an average of 6% per year.

(6) The demand for UTGHR is linked to availability as long as it remains scarce or highly priced potential demand will not be converted to actual demand.

(7) Inevitably, both environmental and energy cost pressures on steelmakers will continue to rise.

(8) Capital availability for steelmakers will continue to be limited as long as the industry demonstrates that it is unable to generate satisfactory returns on investment over the long run.

Given the above situation analysis, the question naturally follows: Is there a process available to the industry and its customers, which will overcome the limitations of the present technologies and simultaneously benefit customers, producers and capital sources?

The answer lies in twin-roll thin strip casting. The inherent nature of the twin-roll casting process has the following characteristics:

(1) The unit size for a CASTRIP plant is about 500000 tonnes per year. Units will fit easily into geographic regions with relatively small demand. At the same time, it is unlikely that adoption of twin-roll casting technology will add to net capacity in the industry—replacement of inefficient *slab casters/breakdown mills/Steckel mills/planetary mills/hot strip mills/cold strip mills* is much more likely.

(2) The CASTRIP process is a natural *complement* to existing *thin slab/thick slab continuous rolling processes*. The CASTRIP process becomes more efficient and productive with decreasing cast thickness. Although the upper technical thickness limit for twin-roll casters has not yet been established (experiments have been successful up to at least 3.5mm) it appears likely that the upper economic limit is about 2.5mm. Hence, the most efficient production range for a CASTRIP plant exactly complements the most economic thickness range available from a hot strip mill.

(3) The Nucor CASTRIP facility will produce commercial quality strip between 0.7mm and 2.0mm in thickness and in widths of from 1000mm to 1500mm (though the equipment is designed to enable production of maximum 2000mm widths). Oxide levels on the strip will be low enough and surface quality high enough that for many uses pickling and *cold rolling* will not be required.

(4) From *ladle* to *finished coil*, the CASTRIP process uses between 80% and 90% less energy than conventional processes, with the accompanying decrease in the production of *greenhouse gases*. For this reason, the United States Department of Energy has identified strip casting as a key technology in its "Technology Roadmap".

(5) The *metallurgy* of twin-roll strip casting is quite unlike conventional ferrous casting and rolling. An unexpected outcome of the process is the ability to achieve significantly different coil properties with a single chemistry of molten steel. This will, in turn, reduce *inventory* significantly.

(6) Financial returns from the installation of CASTRIP facilities appear very attractive. These returns remain to be confirmed by the actual experience of the Nucor CASTRIP facility.

Construction of the world's first commercial strip caster for carbon and stainless steels is complete. When *commissioned*, the CASTRIP facility is expected to have an annual capacity of 500000 tonnes of hot-rolled coil below 2mm in thickness. The production of flat-rolled products utilizing the CASTRIP process has many advantages over conventional casting and rolling technologies, including lower capital and *operating costs*, reduced energy usage and emissions, thinner and higher value products. Further, because of the natural *proclivity* of the CASTRIP process to produce lighter gauges (1.5mm and less) and excellent surface quality, CASTRIP products can be substituted for *cold-rolled sheet* in many applications and will likely create a new *product category*, already termed *UCS*, for flat-rolled sheet products. However, for hot-rolled carbon steel products greater than about 2.5mm it is likely that conventional thick and thin slab casters followed by hot strip mills will remain the preferred processes, for at least the near term.

Words and Expressions

twin-roll strip casting 双辊带钢铸轧
near their final shape 近终形
pilot plant 实验工厂
austenitic stainless steel 奥氏体不锈钢
promising 有希望的
304 stainless coils 304 不锈钢卷
carbon steel 碳素钢
low-carbon heat 低碳钢炉次
ladle 钢包
order 订单
EAF (electric arc furnace) 电弧炉
culminating with 最后的
cold-rolled 冷轧的
metallic-coated 金属涂层的
roofing product 屋顶板
subsidiary 分厂、分部
CASTRIP 铸轧带钢
pipe and tube 管线钢
molten steel 钢液, 钢水
multiheat 多炉次
Castrip LLC (Limited Liability Company)
Castrip 有限责任公司
licensee 执照, 许可
mold 结晶器
meniscus 弯月面
roll nip 辊端

pinch point 夹送点（出口点）
ceramic 陶瓷
mathematical modeling 数学模型
process metallurgy 过程冶金
breakthrough 突破
metal delivery 钢水浇铸
edge containment 侧封
roll distortion 铸辊变形
refractory 耐火材料
conventional casting 常规连铸
slab casting 板坯连铸
shell 凝固外壳
mold powder 结晶器保护渣
mold oscillation 结晶器振动
heat transfer 热传递
melt pool 熔池
side dam material 侧封材料
premature freezing 过早的凝固
triggering 触发
thermal stress 热应力
heat flux 热流量
crown 凸度
strip profile 带钢横断面
active oxygen 溶解在钢中的氧
alumina graphite nozzle 铝碳质水口
CO bubble 一氧化碳气泡

start-up 投产
ladle metallurgy furnace (LMF)
　钢包冶炼（炉外精炼）
ladle turret 钢包转运台
slab caster 板坯连铸机
reheat furnace 再加热炉
hot strip mill 热带轧机
tundish 中间包
dummy bar 引锭杆
pinch roll 夹送辊
hot rolling stand 热轧机座
scale 氧化铁皮
breakdown mill 初轧机
Steckel mill Steckel（斯泰克尔）轧机
planetary mill 行星轧机
hot cold strip mill 热/冷带轧机
complement 补充
thin slab/thick slab 薄/厚板坯
continuous rolling process
　连续轧制过程
cold rolling 冷轧
finished coil 成品钢卷
greenhouse gas 温室气体
metallurgy 冶金
inventory 库存
commission 投产
operating cost 生产成本
proclivity 倾向
cold-rolled sheet 冷轧薄板
product category 产品种类
UCS 铸轧带钢

8 Metal Casting Processes

Casting is one of the earliest metal shaping methods known to human being. It generally means *pouring* molten metal into a *refractory mould* with a *cavity* of the shape to be made, and allowing it to solidify. When solidified, the desired metal object is taken out from the refractory mould either by breaking the mould or taking the mould apart. The solidified object is called casting. This process is also called *founding*.

8.1 History

The discovery of the casting process was probably around 3500 BC in *Mesopotamia*. In many parts of the world during that period, copper *axes* and other flat objects were made in *open moulds* made of stone or *baked clay*. These moulds are essentially in single piece. But in later periods, when round objects were required to be made, the mould was split into two or more parts to *facilitate* the withdrawal of the round objects.

The bronze age (2000 BC) brought far more refinement into casting process. For the first time perhaps, *core* for making *hollow sockets* in the objects was invented. These cores were made of baked clay. Also the *lost wax* process was extensively used for making *ornaments* and fine work.

Casting technology has been greatly improved by Chinese from around 1500 BC. Before that there is no evidence of any casting activity found in China. They do not appear to have been greatly familiar with the lost wax process nor used it extensively but instead specialized in the *multi-piece moulds* for making highly intricate jobs. They spent a lot of time in perfecting the mould to the last detail so that hardly any finishing work was required on the casting made from the moulds. They had probably made piece moulds containing carefully fitted pieces numbering thirty or more, In fact, many such moulds have been unearthed during the *archaeological excavations* in various parts of China.

Indus valley civilization is also known for their extensive use of casting of copper and bronze for ornaments, weapons, tools and *utensils*. But there was not much of improvement in the technology. From the various objects and *figurines* that were excavated from the Indus valley sites, they appear to have been familiar with all the known casting methods such as open mould, piece mould and the lost wax process.

Though India could be credited with the invention of *crucible steel*, not much of iron founding was evident in India. There is evidence that iron founding had started around 1000 BC in Syria and Persia. It appears that iron casting technology in India has been in use from the times of the invasion of Alexander the Great, around 300 BC.

The famous iron pillar presently located near the Qutab Minar in Delhi is an example of the metallurgical skills of ancient Indians. It is 7.2m long and is made of pure *malleable iron*. This is assumed to be of the period of Chandragupta II (375~413 AD) of Gupta dynasty. The rate of rusting of this pillar which stands outside is practically zero and even the buried portion is rusting at extreme slow rate. This must have been first cast and then *hammered* to the final shape.

8.2 Advantages and Limitations

Casting process is extensively used in manufacturing because of its many advantages. Molten material flows into any small section in the *mould cavity* and as such any intricate shapes internal or external can be made with the casting process. It is possible to cast practically any material be it *ferrous* or *non-ferrous*. Further, the necessary tools required for casting moulds are very simple and inexpensive. As a result, for trial production or production of a small lot, it is an ideal method. It is possible in casting process, to place the amount of material where exactly required. As a result, weight reduction in design can be achieved. Castings are generally cooled uniformly from all sides and therefore they are expected to have no *directional properties*. There are certain metals and alloys which can only be processed by the casting and not by any other process like *forging* because of the metallurgical considerations. Casting of any size and weight, even up to 200 tons can be made.

However the *dimensional accuracy* and *surface finish* achieved by normal *sand casting* process would not be adequate for final application in many cases. To take these cases into consideration, some special casting processes such as *die casting* have been developed. Also the sand casting process is labour intensive to some extent and therefore many improvements are aimed at it such as *machine moulding* and foundry mechanisation. With some materials it is often difficult to remove *defects* arising out of the moisture present in sand castings.

8.3 Applications

Typical applications of sand casting process are *cylinder blocks*, *liners*, *machine tool beds*, pistons, *piston rings*, mill rolls, wheels, housings, water supply pipes and specials, and bells.

8.4 Casting Terms

In the following chapters, the details of sand casting process which represents the basic process of casting would be seen. Before going into the details of the process, defining a number of casting vocabulary words would be appropriate. Reference may please be made to Fig. 8-1.

Pattern: Pattern is a *replica* of the final object to be made with some modifications. The mould cavity is made with the help of the pattern.

Parting line: This is the dividing line between the two moulding *flasks* that makes up the sand

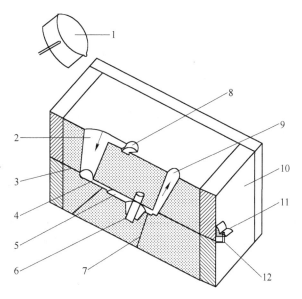

Fig. 8-1 Cross section of a sand mould

1—ladle; 2—spruce; 3—parting line; 4—runner; 5—mould cavity; 6—core;
7—venthole; 8—pouring basin; 9—riser; 10—cope; 11—drag; 12—locating pin

mould. In split pattern it is also the dividing line between the two halves of the pattern.

Core: It is used for making hollow cavities in castings.

Pouring basin: A small funnel shaped cavity at the top of the mould into which the molten metal is poured.

Sprue: The passage through which the molten metal from the pouring basin reaches the mould cavity. In many cases it controls the flow of metal into the mould.

Runner: The passageways in the parting plane through which molten metal flow is regulated before they reach the mould cavity.

Gate: The actual entry point through which molten metal enters mould cavity.

Riser: It is a reservoir of molten metal provided in the casting so that hot metal can flow back into the mould cavity when there is a reduction in volume of metal due to solidification.

8.5 Sand Mould Making Procedure

The procedure for making a typical sand mould is described in the following steps.

First a *bottom board* is placed either on the *moulding platform* or on the floor, making the surface even. The *drag moulding flask* is kept upside down on the bottom board along with the drag part of the pattern at the centre of the flask on the board. There should be enough *clearance* between the pattern and the walls of the flask which should be of the order of 50mm to 100mm. Dry *facing sand* is sprinkled over the board and pattern to provide a non-*sticky* layer. Freshly prepared *moulding sand* of requisite quality is now poured into the drag and on the pattern to a thickness of

30mm to 50mm. Rest of the drag flask is completely filled with the *backup sand* and uniformly rammed to compact the sand. The ramming of sand should be done properly so as not to compact it too hard, which makes the escape of gases difficult, nor too loose so that mould would not have enough strength. After the ramming is over, the excess sand in the flask is completely scraped using a *flat bar* to the level of the *flask edges*.

Now, with a vent wire which is a wire of 1mm to 2mm diameter with a pointed end, *vent holes* are made in the drag to the full depth of the flask as well as to the pattern to facilitate the removal of gases during casting solidification. This completes the preparation of the drag.

The finished drag flask is now rolled over to the bottom board exposing the pattern. Using a *slick*, the edges of sand around the pattern is repaired and cope half of the pattern is placed over the drag pattern, *aligning* it with the help of *dowel pins*. The cope flask on top of the drag is located aligning again with the help of the pins. The dry parting sand is sprinkled all over the drag and on the pattern.

A sprue pin for making the sprue passage is located at a small distance of about 50mm from the pattern. Also a riser pin if required, is kept at an appropriate place and freshly prepared moulding sand similar to that of the drag along with the backing sand is sprinkled. The sand is thoroughly rammed, excess sand scraped and vent holes are made all over in the *cope* as in the *drag*.

The sprue pin and the riser pin are carefully withdrawn from the flask. Later the pouring basin is cut near the top of the sprue. The cope is separated from the drag and any loose sand on the cope and drag interface of the drag is blown off with the help of *bellows*. Now the cope and the drag pattern halves are withdrawn by using the *draw spikes* and *rapping* the pattern all around to slightly enlarge the mould cavity so that the mould walls are not *spoiled* by the withdrawing pattern. The runners and the gates are cut in the mould carefully without spoiling the mould. Any excess or loose sand found in the runners and mould cavity is blown away using the bellows. Now the facing sand in the form of a paste is applied all over the mould cavity and the runners which would give the finished casting a good surface finish.

A dry sand core is prepared using a *core box*. After suitable *baking*, it is placed in the mould cavity as shown in Fig. 8-2. The cope is replaced on the drag taking care of the alignment of the two by means of the pins. The mould now, as shown in Fig. 8-2 is ready for pouring.

8.6 Melting Equipment for Non-ferrous Foundries

In non-ferrous foundries, the capital and energy costs of electric furnaces are generally little higher than those for fuel fired furnaces.

However, when total melting costs are considered the economics often favor electric furnaces, because improvements can be expected in respect of: temperature control, reject castings, metal losses, labour costs, flexibility in choice of charge materials, and contamination from fuels.

Recent developments in the non-ferrous industry have centred on two different types of electric furnace. *Medium frequency coreless furnaces* have been installed in a number of foundries producing

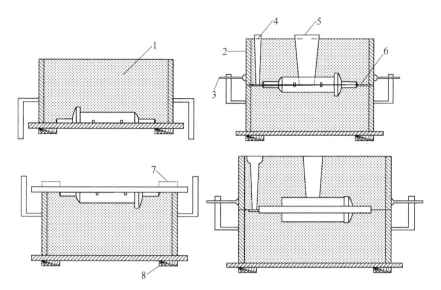

Fig. 8-2 Sand mould making procedure
1—moulding sand; 2—cope task; 3—lug with hole for alignment pin; 4—sprue pin;
5—riser pin; 6—parting surface; 7—mould board (will now be removed); 8—bottom board

copper alloys and *resistance heated crucible furnaces* have become increasingly popular in aluminum foundries.

Most of copper alloy foundries are small, often producing only a few tons of castings per week with a frequent need to produce a wide variety of alloys, A medium frequency coreless furnace will give the foundry this facility and will also give the additional benefits of: lower melting losses, more rapid melting, better quality castings and an improved environment.

The *induction heated* "pop up" or "push out" crucible furnace is gaining in popularity and a number of them are now operating very successfully. The main feature of this furnace is that each alloy can be melted in its own crucible so preventing cross contamination between melts. Furthermore, the high power inputs possible with frequencies in the range 500~10000Hz give very rapid melting. For example, a typical batch of copper alloy can be melted in as little as 20min. The crucible capacity tends to be relatively small, usually 100kg or less, since it has to be man handled from the furnace coil, although larger crucible are available. The installation is normally equipped with two induction coils and powered from either a static or rotary frequency converter. It is usual to provide a change-over switch so that, when a melt is complete, power can be immediately switched to the other coils to melt the next batch of metal.

Renewed interest is being shown by small aluminum foundries in resistance heated by small aluminum foundries in resistance heated crucibles and recently a number of foundries have replaced oil fired crucibles. The new design of resistance heated crucible is comparable in cost with a similar fuel fired furnace of equivalent melting rate and capacity. The main feature of the resistance exhaust gases which considerably improves the foundry environment. Other benefit are longer crucible life and lower running costs due to the higher standard of thermal insulation used in the furnaces.

The choice of melting and holding equipment for both ferrous and non-ferrous foundries must be based on a detailed evaluation of all the factors contributing to the cost of producing molten metal. Growing numbers of foundries are finding that the evaluation will prove the case for electricity which brings the added advantages of improved metallurgical control.

8.7 Fluidity and Pouring

Fluidity as used in the cast metals industry is quite different from the term defined in physical chemistry. In metal casting, it is the ability to fill a mold. To provide a measure of fluidity, a variety of evaluative methods have been developed, where fluidity is measured by pouring a standard mold to provide a good indication of metal flow. Usually a long, thin casting is poured in the form of a *spiral*-the length of the spiral serving as the measure of fluidity.

Both metal and mold characteristics are involved in determining metal fluidity. Two major metallurgical factors have a marked influence on metal fluidity-metal composition with particular emphasis on its relation to the freezing process and superheat. Other factors, such as metal viscosity and surface tension also may be important factors contributing to a metal's fluidity.

With regard to superheat, as a metal is heated to a high temperature, it will have a longer period in the mold during which it is liquid, and, therefore, will flow farther than a metal that is not so highly heated. Absorbed gases, surface oxide films and suspended *inclusions*, can materially affect the degree of fluidity.

Changes in metal composition alter the solidification pattern of a metal or alloy. A long solidification range, where the metal is in a mushy condition, will tend to restrict fluidity, while a short solidification range, as exists with pure metals, does not have the tendency to restrict the flow of the still liquid interior metal.

There are many known modifications to the spiral fluidity test. While the spiral has been used for both ferrous and non-ferrous metals, the Briggs-Gezelius non-spiral fluidity test is used, extensively with ferrous metal-particularly steels and high alloys.

Pouring is the process by which molten metal is made to enter the mould in foundry work. The metal in this state is carried in a ladle to the mould and poured in various ways, according to the size and type of pouring vessel. The moulds are filled as rapidly as possible, and after a series has been cast, the ladle is taken back to the start of a fresh run to provide additional hot metal to the risers.

The molten metal comes to the top of the risers, whereupon pouring stops, and a layer of exothermic insulating powder is placed on the surface of the metal to prevent it solidifying. It is often advantageous to deposit an insulating moulding mixture round the risers. Various addition are made to the molten metal before it is poured, e. g. oxidizing agents, *grain refining additions*, metals to produce desired alloy compositions, *deoxidizers*, etc.

Before the metal passes into the mould cavity from the ladle it is received by a pouring basin or bowel when the casting moulds are small and made from *green sand*. This basin, made from dry sand, is placed over the main sprue when pouring begins, but may occasionally be formed in the top half of the mould. It normally embodies a dam that allows a proportion of the molten metal to build up in the basin in advance of its journey into the mould. In this way a more uniform flow is achieved. A skim core may also be embodied, consisting of a section formed in the upper part of the basin to skim off surface matter that must not enter the mould.

The purpose of the pouring basin is to reduce the pouring velocity and impact that mould occur if the stream fell directly from ladle into mould. It also enables a good deal of dross or dirt to be removed, whether or not a skim is used.

Words and Expressions

casting 铸造，铸件
pour 浇注
refractory 难熔化的
mould 铸型
cavity 空腔
founding 铸造
Mesopotamia
    美索不达米亚（西南亚地区）
axes 轴
open mould
    暴露的铸型，即没有盖箱的铸型
baked clay 烘烤过的黏土
facilitate 使…容易
core 芯子
hollow socket 凹孔
lost wax 失蜡（铸造）
ornament 装饰品
multi-piece mould
    分模，即由多块模型拼凑成的铸型
archaeological excavation 考古发掘
Indus valley 印度西北部河谷
Utensil 器具
Figurine 小雕像
crucible steel 坩埚钢
malleable iron 可锻铸铁
hammered 锻打
mould cavity 型腔

ferrous 含铁的，即黑色（金属）
non-ferrous 不含铁的，即有色（金属）
directional properties 性能的方向性
forging 锻造
dimensional accuracy 尺寸精度
surface finish 表面光洁度
die casting 压力铸造
sand casting 砂型铸造
machine mould 机械制模（造型）
defects 缺陷
cylinder blocks 汽缸体
liner 垫圈
machine tool beds 机床床身
piston ring 活塞环
pattern 模型
replica 复制品
parting line 分型线
flask 砂箱
cope 上箱
drag 下箱（=drag moulding flask）
pouring basin 浇口杯
sprue 直浇道
runner 横浇道
gate 内浇道
riser 冒口
bottom board 底平板
moulding platform 造型台

clearance 吃砂量
facing sand 面砂
sticky 黏性的
moulding sand 型砂
backup sand 背砂
flat bar 扁平棒
flask edge 砂箱边框
vent hole 通气孔
slick （造型用）修磨工具
aligning 校正
dowel pins 销钉
bellows （吹风用的）皮老虎
draw spike 起模针
rapping 轻击

spoil 弄坏
core box 芯盒
baking 焙烧
medium frequency coreless furnace
　中频感应无芯电炉
resistance heated crucible furnace
　电阻加热坩埚炉
induction heat 感应加热
spiral 螺旋形
inclusion 夹杂物
grain refining addition
　晶粒细化剂
deoxidizer 还原剂
green sand 造型用砂

9 Forging

9.1 Classification of Forging Process

Forging is the working of metal into a useful shape by hammering or *pressing*. It is the oldest of the metalworking arts. Having its origin with the *primitive blacksmith* of Biblical times. The develop of machinery to replace the arm of the smith occurred early during the Industrial Revolution. Today there is a wide variety of forging machinery, which is capable of making parts ranging in size form a *bolt* to a *turbine rotor* or an entire airplane wing.

Most forging operations are carried out hot, although certain metals may be cold-forged. Two major classes of equipment are used for forging operations. The *forging hammer*, or *drop hammer*, delivers rapid *impact blows* to the surface of the metal, while the *forging press* subjects the metal to a slow-speed compressive force.

The two broad categories of forging processes are *open-die forging* and *closed-die forging*. Open-die forging is carried out between flat dies or dies of very simple shape. The process is used mostly for large objects or when the number of parts produced is small. Often open-die forging is used to preform the workpiece for close-die forging. In closed-die forging the workpiece is deformed between two die halves which carry the *impressions* of the desired final shape. The workpiece is deformed under high pressure in a closed cavity, and thus precision forgings with close *dimensional tolerances* can be produced.

The simplest open-die forging operation is the *upsetting* of a cylindrical billet between two flat dies. The compression test is a small-scale prototype of this process. As the metal flows laterally between the advancing die surfaces. There is less deformation at the die interfaces because of the forces than at the midheight plane. Thus, the side of the upset cylinder becomes barreled. As a general rule, metal will flow most easily toward the nearest *free surface* because this represents the lowest frictional path.

The effect of friction in restraining metal flow is used to produce shapes with simple dies. *Edging* dies are used to shape the ends of the bars and to gather metal. As is shown in Fig. 9-1(a) and Fig. 9-1(b), the metal is confined by the die from flowing in the horizontal direction but it is free to flow laterally to fill the die. *Fullering* is used to reduce the cross-sectional area of a portion of the stock. The metal flow is outward and away form the center of the fullering die (Fig. 9-1 (c)). An example of the use of this type of operation would be in the forging of a connecting rod for an internal-combustion engine. The reduction in cross section of the work with concurrent increase in length is called drawing down, or drawing out (Fig. 9-1(d)). If the drawing-down operation is carried out with concave dies (Fig. 9-1(e)) so as to produce a bar of smaller diameter, it is called *swaging*. Other operations which can be achieved by forging are bending, *twisting*, *extrusion*, *piercing* (Fig. 9-1(f)), *punching* (Fig. 9-1(g)), and *indenting*.

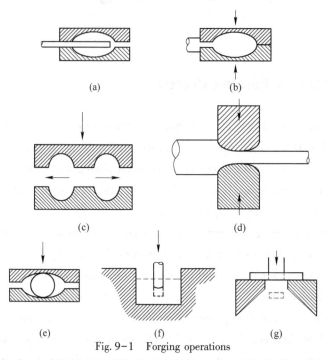

Fig. 9-1 Forging operations
(a), (b) edging; (c) fullering; (d) drawing; (e) swaging; (f) piercing; (g) punching

Closed-die forging uses carefully machined matching *die blocks* to produce forgings to close dimensional tolerances. Large production runs are generally required to justify the expensive dies. In closed-die forging the forging billet is usually first fullered and edged to place the metal in the correct places for subsequent forging. The preshaped billet is then placed in the cavity of the *blocking die* and rough-forged to close to the final shape. The greatest change in the shape of the metal usually occurs in this step. It is then transferred to the *finishing die*, where it is forged to final shape and dimensions, Usually the *blocking cavity* and the *finishing cavity* are machined into the same die block. Fullering and edging impressions are often placed on the edges of the die block. For complex shapes more than one preforming or blocking operation is required to achieve a gradual flow of metal from the initial billet to the complex final shape.

It is important to use enough metal in the forging billet so that the die *cavity* is completely filled. Because it is difficult to put just the right amount of metal in the correct places during fullering and edging, it is customary to use a slight excess of metal. When the dies come together for the finishing step, the excess metal squirts out of the cavity as a thin *ribbon* of metal called *flash*. In order to prevent the formation of a very wide flash, a ridge, known as a flash *gutter*, is usually provided (Fig. 9-2). The final step in making a closed-die forging is the removal of the flash with a *trimming die*.

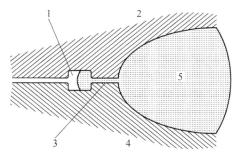

Fig. 9-2　Sectional view through closed-die forging
1—flash gutter; 2—upper die; 3—flash; 4—lower die; 5—forging

Because of the flash, the term closed-die forging is a bit of a misnomer, and a better description for the process would be *impression-die forging*. The flash serves two purposes. As describes above, it acts as a "safety valve" for excess metal in the closed-die cavity. Of more importance, the flash regulates the escape of metal, and thus the thin flash greatly increases the flow resistance of the system so that the pressure builds up to high values to ensure that metal fills all recesses of the die cavity. Fig. 9-3 shows a typical curve of forging load vs. die advance (press stroke) for a closed-die forging process. The trick in designing the flash is to adjust its dimensions so that the extrusion of metal through the narrow flash opening is more difficult than the filling of the most *intricate detail* in the die. But this must not be done to excess so as to create very high forging loads with attendant problems of die wear and breakage. The ideal is to design for the minimum flash needed to do the job. Forging pressure increases with decreasing flash thickness and increasing flash land width.

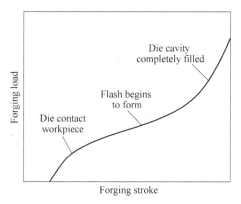

Fig. 9-3　Typical curve of forging load vs. stroke for closed-die forging

The metal flow is greatly influenced by the part geometry. Spherical or blocklike shapes are easiest to forge in impression dies. Shapes with thin and long sections or *projections* (*ribs* and *webs*) are more difficult because they have higher surface area per unit volume, and therefore friction and temperature effects are enhanced. It is particularly difficult to produce parts with sharp fillets, wide thin webs, and high ribs. Moreover, forging dies must be tapered to facilitate removal of the finished piece. This *draft allowance* is approximately 5° for steel forgings.

9.2 Forging Equipment

Forging equipment may be classified with respect to the principle of operation. In forging hammers the force is supplied by a *falling weight* or ram. These are energy-restricted machines since the deformation results from *dissipating* the kinetic energy of the ram. Mechanical forging presses are *stroke*-restricted machines since the length of the press stroke and the available load at various positions of the stroke represent their capability. *Hydraulic presses* are load-restricted machines since their capability for carrying out a forming operation is limited chiefly by the maximum *load capacity*. Each of these classes forging equipment needs to be examined with respect to its load and energy characteristics, its time-dependent characteristics, and its capability for producing parts to dimension with high accuracy.

To successfully complete a forging operation, the *available machine load* must exceed the required load at any point in the process and the *available machine energy* must exceed the energy required by the process for the entire stroke. The most important characteristic of any machine is the number of strokes per minute, for this determines the production rate. The velocity under pressure v_p is the velocity of the machine slide under load. This variable determines the strain rate (which influences the *flow stress*) and the contact time under pressure t_p. The contact time is the time that the workpiece remains in the die under load. Since heat transfer between the hotter forging and the cooler dies is so effective when the interface is under high pressure, die wear increases with t_p. Dimensional accuracy of the parts produced by a forging machine is directly related to the *stiffness* of the equipment. In general terms, stiffness is increased by using larger components in the construction of the machine, and this is directly related to increased equipment cost. A particularly critical problem is the tilting of the ram under load.

The most commonly used piece of forging equipment is the forging hammer. The two basic types of hammers are the *board hammer* and the *power hammer* (Fig. 9-4). In the board hammer the upper die and ram are raised by friction rolls gripping the board. When the board is released, the ram falls under the influence of gravity to produce the *blow energy*. The board is immediately raised again for another blow. Forging under a hammer usually is done with repeated blows. Hammers can strike between 60 and 150 blows per minute depending on size and capacity. The energy supplied by the blow is equal to the *potential energy* due to the weight of the ram and the height of the fall. Forging hammers are rated by the weight of the ram. However, since the hammer is an energy-restricted machine, in which the deformation proceeds until the total kinetic energy is dissipated by plastic deformation of the workpiece or elastic deformation of the dies and machine, it is more correct to rate these machines in terms of energy delivered.

9.3 Open-die Forging

Open-die forging typically deals with large, relatively simple shapes that are formed between sim-

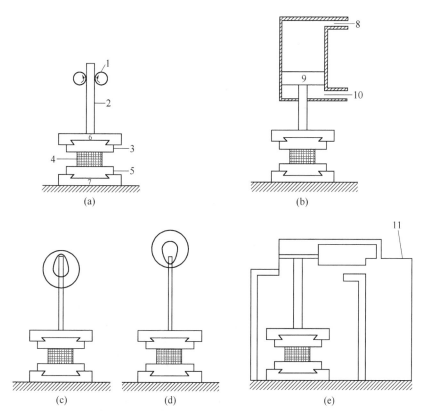

Fig. 9-4 Schematic drawings of forging equipment
(a) board hammer; (b) steam hammer; (c) *crank press* (top strake); (d) crank press (bottom strake); (e) hydraulic press
1—roll; 2—board; 3—upper die; 4—work; 5—lower die; 6—ram; 7—*anvil*; 8—down; 9—piston; 10—up; 11—intensifier

ple dies in a large hydraulic press or power hammer. Examples of parts of made in open-die forging are ship *propeller shafts*, rings, *gun tubes* and pressure vessels. Since the workpiece is usually larger than the tool, at any point in time deformation is confined to a small portion of the workpiece. The chief mode of deformation is compression, accompanied by *considerable spreading* in the *lateral directions*.

Probably the simplest open-die forging operation is *cogging* a billet between flat tools to reduce the *cross-sectional* area, usually without changing the final shape of the cross section. Fig. 9-5 helps to define the *nomenclature* in dealing with spread in cogging. Tomlinson and Stringer defined a coefficient of spread S

$$S = \frac{\text{width } elongation}{\text{Thickness contraction}} = \frac{\ln(w_1/w_0)}{\ln(h_0/h_1)} \tag{9-1}$$

Because of barreling of the bar, it is difficult to measure the width *natural strain*, but the increase in length can be measured accurately. Using the constancy of volume *relationship*, we can write

$$\frac{h_1 w_1 l_1}{h_0 w_0 l_0} = 1 \tag{9-2}$$

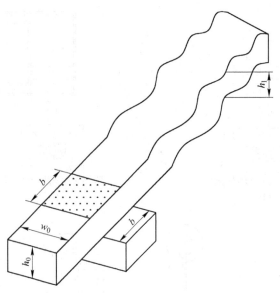

Fig. 9-5 Cogging operation in open die forging. Shaped area shows where contact would occur between wokpiece and upper die

or
$$\ln(h_1/h_0) + \ln(w_1/w_0) + \ln(l_1/l_0) = 0$$

Substituting into Eq. (9-1) gives the coefficient of elongation

$$1 - S = \frac{\text{width elongation}}{\text{Thickness contraction}} = \frac{\ln(l_1/l_0)}{\ln(h_0/h_1)} \tag{9-3}$$

If $S=1$, then all of the deformation could manifest itself as spread, while if $S=0$, all of the deformation would go into elongation. It was found that S depended chiefly on the *bite* ratio $\dfrac{b}{w_0}$ according to

$$S = \frac{b/w_0}{1 + b/w_0} \tag{9-4}$$

Eq. (9-1) is often expressed in terms of the "spread law"

$$\beta = \left(\frac{1}{\gamma}\right)^S \tag{9-5}$$

where
β = *spread ratio* = w_1/w_0

γ = squeeze ratio = h_1/h_0

There are certain limiting ranges of these variables, which must be considered. Since only that part of the surface under the bite is being deformed at any one time, there is danger of causing surface laps at the step separating the forged from the unforged porting of the workpiece. For a given geometry of tooling there will be a *critical deformation*, which will produce laps. Wistreich and Shutt recommend that the squeeze ratio h_0/h_1 should not exceed 1.3. Since open-die forging deformed through to the center. It is recommended that the bite ratio b/h should not be less than $\dfrac{1}{3}$ to minimize *inhomogeneous deformation*. Using these criteria Wistreich and Shutt developed optimization techniques for selecting the forging schedule from the thousands of possible combinations, which would require the least number of steps.

The load required to forge a flat section in open dies may be estimated by

$$p = \overline{\sigma}AC \tag{9-6}$$

where C is a constraint factor to allow for inhomogeneous deformation. It will be recalled that *deformation resistance* increases with $\Delta = h/L$. Hill Constructed *slipline fields* for forging with various conditions of Δ and the results can be summarized by the relation $C = 0.8 + 0.2h/b = 0.8 + 0.2\Delta$.

9.4 Close-die Forging

The description of the closed-die forging process in Sec. 9.2 emphasized the important role of the flash in controlling die fill and in creating high forging loads. Usually the deformation in closed-die forging is very complex and the design of the intermediate steps to make a final precision part requires considerable experience and skill. Overall success of the forging operation requires an understanding of the flow stress of the material, the frictional conditions and the flow of the material in order to develop the optimum geometry for the dies. A special problem in closed-die forging is preventing rapid cooling of the workpiece by the colder dies. Toward this end *isothermal forging* in heated *superalloy* dies is being practiced with difficulty to forge aerospace materials.

An important step toward rationalizing the design of closed-die forgings is the classification of the shapes commonly produced by this process. The degree of difficulty increases as the geometry moves down and toward the right in this illustration. Roughly 70% of forgings fall in the third shape class with one dimension significantly longer than the other two, although this classification system is useful in cost estimating and designing preforming steps, it is not quantitative. More quantitative methods of defining the shape difficulty factor have been developed.

The design of a part for production by closed-die forging involves the prediction of:
(1) Workpiece volume and weight;
(2) Number of preforming steps and their configuration;
(3) Flash dimensions in preforming and finishing dies;
(4) The load and energy requirements for each forging operation.

Steps (1) and (3) have been discussed in detail by Atlan and Henning and the procedures reduce to a working computer program that is applicable to axisymmetric shapes.

Preform design is the most difficult and critical step in forging design. Proper preform design assures defect-free flow, complete die fill, and minimum flash loss. Success here depends on a thorough understanding of the metal flow during forging. Although metal flow consists only of two basic types extrusion (flow parallel to the direction of the die motion) and upsetting (flow perpendicular to the direction of die motion), in most forgings both types of flow occur simultaneously. An important step in understanding metal flow is to identify the neutral to the die motion.

In designing a preform it is usual practice to take key cross sections through the forging and design the preform on the basis of the metal flow. Some general considerations are:

(1) The area at each cross section along the length must equal the area in the finished cross section plus the flash.

(2) All concave radii on the preform should be larger than the radii on the final forged part.

(3) The cross section of the preform should be higher and narrower than the final cross section, so as to *accentuate* upsetting flow and minimize extrusion flow.

Ideally, flow in the finishing step should be lateral toward the *die cavity* without additional shear at the die – workpiece interface. This type of flow minimizes friction, forging load, and die wear. An example of the use of basic principles in preform design has been given by Akgerman, Becker, and Altan.

A milestone in metalworking is the use of computer-aided design (CAD) in establishing the proper design for preforming and finishing dies in closed-die forging. CAD has been applied to rib-web type *airframe* forgings and to *airfoil shapes*, but it can be applied to any class shape for which there is a suitable volume of parts to justify the development work. The power of the CAD system can be appreciated from Fig. 9 – 6. Starting with a drawing of the final part, the CAD system defines this geometry in terms of points, planes, cylinders, and other regular geometric shapes using the APT computer language. APT is a specialized computer language for describing geometric changes produced in metal cutting that is at the heart of numerical controlled (N/C) machining. Next, the coordinates of the various cross sections of the forging are determined, and they are used to preform design calculations to establish such factors as the location of the neutral surface, the shape difficulty factor, the cross-sectional area and volume, the flash geometry, and the stresses, the loads, and the center of loading. An important aspect of this system is that it takes the part geometry and flash dimensions and generates the N/C tape for machining the *electrodes* in the sinking of the finishing dies by *electric discharge machining*. Thus, this system also involves computer-aided manufacturing (CAM). CAM is also used to machine the preforming dies.

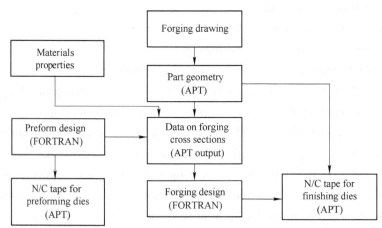

Fig. 9-6 Flow diagram for computer-aided design (CAD) and computer-aided manufacturing (CAM) systems applied to closed-die forging

9.5 Calculation of Forging Loads in Closed-die Forging

The prediction of forging load and pressure in a closed-die forging operation is quite a difficult calculation. There are three general approaches to the problem. The approach used in many *forge shops* is to estimate the forging load required for a new part from information available from previous forging of the same material and similar shape. Slightly more sophistication is found in what might be called the *empirical approach*. Schey has expressed the forging load as

$$P = \overline{\sigma} A_1 C_1 \tag{9-7}$$

where A_1 is cross-sectional area of the forging at the *parting line*, including the flash; C_1 is a *constraint factor* which depends on the complexity of the forging; C_1 has a value of 1.2 to 2.5 for upsetting cylinder between flat dies; C_1 varies from 3 to 8 for closed-die forging of simple shape with flash and from 8 to 12 for more complex shapes.

The third approach is to use the *slab analysis*, suitably modified for the special situations found in closed-die forging. Although this level of analysis does not consider nonuniform deformation, when applied with consideration of the physical situation, it can give good agreement with experimental results.

9.6 Forging Defects

If the deformation during forging is limited to the surface layers, as when light rapid hammer blows are used, the dendritic ingot structure will not be broken down at the interior of the forging. Incomplete forging penetration can readily be detected by macroetching a cross section of the forging. The examination of a deep etch disk for segregation, dendritic structure, and cracks is a standard quality-control procedure with large forgings. To minimize incomplete penetration, forgings of large cross section are usually made on a forging press.

Surface cracking can occur as a result of excessive working of the surface at too low a temperature or as a result of *hot shortness*. A high sulfur concentration in the furnace atmosphere can produce hot shortness in steel and nickel. Cracking at the flash of closed-die forgings is another surface defect, since the crack generally penetrates into the body of the forging when the flash is *trimmed off* (Fig. 9-7(a)). This type of cracking is more prevalent in the thinner the flash in relation to the original thickness of the metal. Flash cracking can be avoided by increasing the flash thickness or by relocating the flash to a less critical region of the forging. It also may be avoided by *hot trimming* or *stress relieving* the forging prior to cold trimming of the flash.

Another common surface defect in closed-die forgings is the *cold shut* or fold (Fig. 9-7(b)). A cold shut is a *discontinuity* produced when two surfaces of metal fold against each other without welding completely. This can happen when metal flows past part of the die cavity that has already been filled or that is only partly filled because the metal failed to fill in due to a sharp corner, excessive *chilling*, or high friction. A common cause of cold shuts is too small a *die radius*.

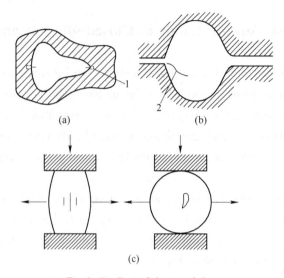

Fig. 9-7 Typical forging defects
(a) cracking at the flash; (b) cold shut or fold; (c) internal cracking due to secondary tensile stresses
1—flash; 2—cold shut

Loose scale or *lubricant residue* that accumulates in deep *recesses* of the die forms scale pockets and causes *underfill*. Incomplete *descaling* of the workpiece results in forged-in scale on the finished part.

Secondary *tensile stresses* can develop during forging, and cracking can thus be produced. *Internal cracks* can develop during the upsetting of a cylinder or a round (Fig. 9-7 (c)), as a result of the *circumferential tensile stresses*. Proper design of the dies, however, can minimize this type of cracking. In order to minimize bulging during upsetting and the development of circumferential tensile stresses, it is usual practice to use concave dies. Internal cracking is less prevalent in closed-die forging because *lateral compressive stresses* are developed by the reaction of the work with the die wall.

The deformation produce by forging results in a certain degree of directionality to the microstructure in which second phases and *inclusions* are oriented parallel to the direction of greatest deformation. When viewed at low magnification, this appears as *flow lines*, or *fiber structure*. The existence of a fiber structure is characteristic of all forgings and is not to be considered as a forging defect. However, the fiber structure results in lower tensile ductility and *fatigue properties* in the direction normal to it (*transverse direction*). To achieve an optimum balance between the ductility in the longitudinal and transverse directions of a forgings, it is often necessary to limit the amount of deformation to 50%~70% reduction in cross section.

9.7 Residual Stresses in Forging

The residual stresses produced in forgings as a result of inhomogeneous deformation are generally small because the deformation is usually carried out well into the hot-working region. However, appreciable residual stresses and *warping* can occur on the quenching of steel forgings in heat treatment.

Special precautions must be observed during the cooling of large steel forgings from the hot-working temperature. Large forgings are subject to the formation of small cracks, or *flakes*, at the center of the cross section flaking is associated with the high hydrogen content usually present in steel ingots of large size, coupled with the presence of residual stresses, in order to guard against the development of high thermal or transformation residual stresses, large forgings are very slowly cooled from the working temperature. This may be accomplished by burying the forging in ashes for periods up to several weeks or in the controlled cooling treatment which is used for hot-rolled railroad rail and certain forgings, by transferring the hot forging to an automatically controlled cooling cycle which brings the forging to a safe temperature in a number of hours. The use of *vacuum-degassed* steel largely eliminates problems with flaking.

Words and Expressions

forge　锻造
press　压力机
primitive blacksmith　原始的手锻
bolt　螺栓
drop hammer　落锤、锻锤、模锻锤
turbine rotor　汽轮机转子
impact blow　冲击作用
forging hammer　锻锤
forging press　锻压机
open-die forging　开式模锻、自由锻
closed-die forging　闭式模锻
dimensional tolerance　尺寸公差
impression　模腔，型腔
upset　镦粗
free surface　自由表面
edging　滚挤，滚压工步
fullering　压槽，压槽工步
swaging　型锻，模锻，旋锻，环锻
twisting　扭转、扭曲
extrusion　挤压
piercing　穿孔、冲孔
punching　冲孔
indenting　压痕，切口

die block　模块，模具坯料，底模
blocking die　预锻模，粗压模
finishing die　精锻模，成型模，终锻模
blocking cavity　预锻模膛
finishing cavity　终锻模膛
cavity　型腔，型槽
ribbon　带
flash　毛边
gutter　槽
trimming die　修边模，冲模
impression-die forging　型腔模锻
intricate detail　模制品
projection　凸出，凸块，凸出部
rib　肋，加强部
web　筋，腹部
draft allowance　拔模斜度
falling weight　下落重量
dissipate　损耗
stroke　行程、冲程
hydraulic press　液压机
load capacity　载荷量，负荷容量
available machine load　有效机械载荷
available machine energy　有效机械能

flow stress 流动应力
board hammer 重力式落锤
anvil 铁砧，砧台，砧，锤砧
stiffness 刚度
power hammer 电锤，动力锤，机动锤
crank press 曲柄压力机
blow energy 打击能量
potential energy 势能
propeller shaft 螺旋桨轴
gun tube 炮管
considerable spreading 相当多的宽展
lateral direction 横向
cogging 压下，开坯
cross-section 横截面
nomenclature 专业术语
elongation 伸长，延伸
natural strain 自然应变
relationship 关系式
substituting 带入
bite 送进量，吃刀量，切入量
critical deformation 临界变形
spread ratio 宽展率，宽展系数
inhomogeneous deformation 不均匀变形
deformation resistance 变形阻力
slipline field 滑移线场
isothermal forging 等温锻造
superalloy 超耐热合金
accentuate 强调
die cavity 模腔
airframe 飞机机架
airfoil shape 翼形
electrode 电极

electric discharge machining 电火花加工
constraint factor 约束系数
forge shop 锻造车间
empirical approach 经验方法
parting line 分型线
slab analysis 主应力法，工程法，切片法
hot shortness 热脆性
trim off 修剪
hot trimming 热切毛边
stress relieving 应力消除
cold shut 冷结
discontinuity 间隔（不连续）
chilling 制冷
die radius 模口圆角半径
loose scale 疏松氧化铁皮
lubricant residue 润滑油渍
recess 凹槽
underfill 未充满
descale 除鳞
tensile stress 拉应力
internal crack 内部裂纹
lateral compressive stress 侧压应力
circumferential tensile stresses 周边拉应力
Inclusion 夹杂物
flow line 流线
fiber structure 纤维组织
fatigue property 疲劳特性
transverse direction 横向
warping 翘曲，扭曲变形
flake 白点
vacuum-degassed 真空除气

10 Sheet-Metal Forming

10.1 Introduction

The ability to produce a variety of shapes from flat sheets of metal at high rates of production has been one of the real technological advances of the 20th century. This transition from hand-forming operations to mass-production methods has been an important factor in the great improvement in the standard of living, which occurred during the period.

In essence, a shape is produced from a *flat blank* by stretching and shrinking the dimensions of all its volume elements in the three mutually perpendicular principal directions. The resulting shape is then the result of the integration of all the local *stretching* and shrinking of the volume elements. Attempts have been made to classify the almost limitless number of shapes which are possible in metal forming into definite categories depending on the contour of the finished part Sachs has classified sheet-metal parts into five categories.

(1) Singly curved parts;
(2) Contoured flanged parts-including parts with *stretch flanges* and *shrink flanges*;
(3) Curved sections;
(4) *Deep-recessed* parts-including cups and boxes with either vertical or sloping walls;
(5) *Shallow-recessed* parts-including dish-shaped, *beaded*, *embossed*, and *corrugated* parts.

Typical examples of these parts are shown in Fig. 10-1. Another classification system, developed in the automotive industry, groups sheet-steel parts into categories depending on the *severity* of the forming operation. Severity of the operation is based on the maximum amount of bending or stretching in the part.

Still another way of classifying sheet-metal forming is by means of specific operations such as bending, *shearing*, *deep drawing*, stretching, *ironing*, etc, Most of these operations have been illustrated briefly in Fig. 10-1, and they will be discussed in considerably greater detail in this chapter.

We should note that unlike the *bulk-forming* deformation processes described in the earlier chapters, *sheet forming* is carried out generally in the plane of the sheet by *tensile forces*. The application of *compressive forces* in the plane of the sheet is avoided because it leads to *buckling*, *folding*, and *wrinkling* of the sheet. While in bulk-forming processes the intention is often to change the thickness or lateral dimensions of the workpiece, in sheet-forming processes decreases in thickness should be avoided because they could lead to *necking* and *failure* Another basic difference between bulk forming and sheet forming is that sheet metals, by their very nature have a high ratio of surface area to thickness.

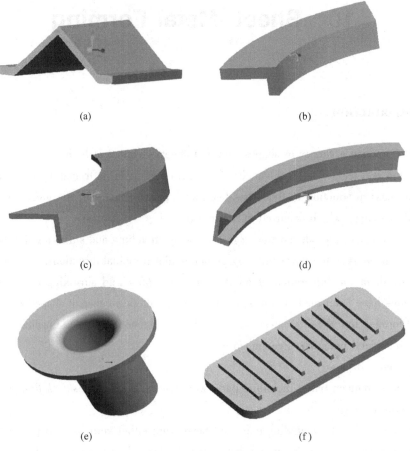

Fig. 10-1 Typical formed shapes
(a) singly curved; (b) stretched flange; (c) shrink flange;
(d) curved sections; (e) deep-drawn cup; (f) beaded section

10.2 Forming Methods

The old method of hand forming of sheet metal is today used primarily as a *finishing operation* to remove wrinkles left by forming machines. In the metal-working industries hand forming is primarily limited to experimental work where only a few identical pieces are required.

Most high-production-volume sheet-metal forming is done on a press, driven by either mechanical or hydraulic action. In mechanical presses energy is generally stored in a flywheel and is transferred to the movable *slide* on the *downstroke* of the press. Mechanical presses are usually quick-acting and have a short stroke, while hydraulic presses are slower-acting but can apply a longer stroke. Presses are usually classified according to the number of slides which can be operated independently of each other. In the *single-action press* there is only one slide, generally operating in the vertical direction. In the *double-action press* there are two slides. The second action ordinarily is used to operate the *hold-down*, which prevents wrinkling in deep drawing. A triple-action press is equipped with two actions above the die and one action below the die.

The basic tools used with a metalworking press are the *punch* and the die. The punch is the convex tool, which mates with the concave die. Generally the punch is the moving element. Because accurate alignment between the punch and die is usually required, it is common practice to mount them permanently in a *subpress*, or *die set*, which can quickly be inserted in the press. An important consideration in tooling for sheet-metal forming is the frequent requirement for a *clamping pressure*, or hold-down, to prevent wrinkling of the sheet as it is being formed. Hold-down can best be provided by a *hold-down ring*, which is actuated by the second action of a double-action press. However, by using mechanical springs or an auxiliary *air cylinder*, hold-down can be provided in a single-action press.

Frequently punches and dies are designed so that successive stages in the forming of the part are carried out in the same die on each stroke of the press. This is known as *progressive forming*. A simple example is a progressive blanking and piercing die to make a plain, flat washer(Fig. 10-2). As the strip is fed from left to right, the hole for the *washer* is first punched and then the washer is *blanked* from the strip. At the same time as the washer is being blanked from the strip, the punch A is *piercing* the hole for the next washer. The *stripper plate* is used to prevent the metal from separating from the die on the up stroke of the punch.

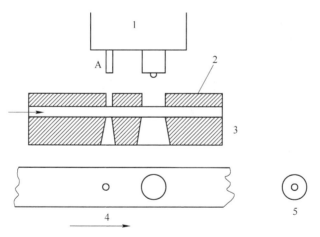

Fig. 10-2 Progressive piercing and blanking die
1—punch; 2—stripper plate; 3—die; 4—strip; 5—washer

Compound dies are designed to perform several operations on the same piece in one stroke of the press. Because of their complexity the dies are costlier and the operations somewhat slower than individual operations. Another strategy is to use *transfer dies*, where the part is moved from station to station within the press for each operation. The die materials depend on the severity of the operation and the required production run. In aircraft work, where production runs are often small, tooling is frequently made from a *zinc-base alloy* called *Kirksite* or from wood or *epoxy re-*

sins. For long die life, however, tool steel is required.

The *press brake* is a single-action press with a very long and narrow bed. The chief purpose of a press brake is to form long, straight bends in pieces such as channels and corrugated sheets. *Roll forming* is another common method of producing bent shapes in long lengths. The roll-forming process is also used to produce thin-wall cylinders from flat sheet.

Rubber hydroforming is a modification of the conventional punch and die in which a pad of rubber or *polyurethane* serves as the die. Rubber forming or the *Guerin process*, is illustrated in Fig. 10-3. A *form block* (punch) is fastened to the bed of a single-action hydraulic press, and a thick blanket of rubber is placed in a retainer box on the upper *platen* of the press. When a blank is placed over the form block and the rubber forced down on the sheet, the rubber transmits a nearly uniform hydrostatic pressure against the sheet. A unit pressure of around 1,500 psi is sufficient for most parts, and higher local pressures can be provided by auxiliary tooling. The Verson-Wheelon process uses a soft *rubber bag* subjected to internal fluid pressure. Because the forming pressure is four to five times greater than in the Guerin process it can be used to form more complicated and deeper shapes. *Rubber forming* is used extensively in the aircraft industry. Shallow flanged parts with stretch flanges are readily produced by this method but shrink flange are limited because the rubber provides little resistance to wrinkling. Another limitation is that the blank tends to move on the form block unless holes for *positioning pins* are provided in the part.

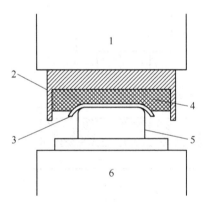

Fig. 10-3 Rubber forming

1—upper platen; 2—*retainer*; 3—blank; 4—rubber pad; 5—form block; 6—press bed

A variety of methods are used to bend or to contour-form straight sections. Cylindrical-and conical-shaped parts are produced with *bending rolls* (Fig. 10-4 (a)). A three-roll bender is not very well suited to preventing buckling in *thin-gage* sheet. Often a fourth roll is placed at the exit to provide an extra adjustment in *curvature*. In three-point loading the maximum *bending moment* is at the midpoint of the span. This localization of strain can result, under certain circumstances, in the *forming limit* being reached at the midpoint before the rest of the part is bent to the proper contour. More uniform deformation along the length of the part is obtained with *wiper-type* equipment. In its simplest form this consists of a sheet which is clamped at one end against a form block: the contour is progressively formed by successive hammer blows, starting near the clamp and moving a short distance toward the free end with each blow. A wiper-type bender is sketched in Fig. 10-4(b). In this case the form block or die has a nonuniform contour so that the *wiper rolls* must be pressed against the block with a uniform pressure supplied by a *hydraulic cylinder*. Still a third method of producing contours is by wrap forming. In *wrap forming* the sheet is compressed against a form block, and at the same time a longitudinal tensile stress is applied to prevent buckling and wrinkling (Fig. 10-4(c)). A simple example of wrap forming is the coiling of a spring around a *mandrel*. The stretch forming of curved sections is a special case of wrap forming.

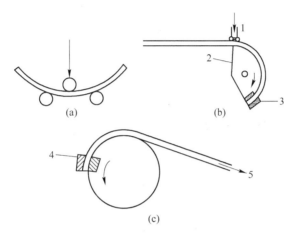

Fig. 10-4 Methods of bending and contouring
(a) three-roll bender; (b) wiper-type benders; (c) wrap forming
1—wiper rolls; 2—form block; 3, 4—clamp; 5—tension

A method of making *tank heads*, television cones, and other deep parts of circulars symmetry is *spinning* (Fig. 10-5(a)). The metal blank is clamped against a form block which is rotated at high speed. The blank is progressively formed against the block, either with a manual tool or by means of small-diameter work rolls. In the spinning process the blank thickness does not change but its diameter is decreased. The *shear-spinning* process. Fig. 10-5(b) is a variant of conventional spinning. In this process the part is reduced according to $t = t_0 \sin\alpha$. This process is also known as *power spinning*, *flow turning*, and *hydrospinning*, It is used for large axisymmetrical conical or *curvilinear* shapes such as *rocket-motor casings* and *missile nose cones*, Still a third variation of spinning is *tube spinning* in which a tube is reduced in wall thickness by spinning on a mandrel. The spinning tool can operate on either the outside or inside diameter of the tube.

Explosive forming is well suited to produce large parts with a relatively low production lot size. The sheet-metal blank is placed over a die cavity and an *explosive charge* is *detonated* in water at an appropriate *standoff distance* from the blank. The shock wave propagating from the explosion serves as a "frictionless punch" to deform the blank.

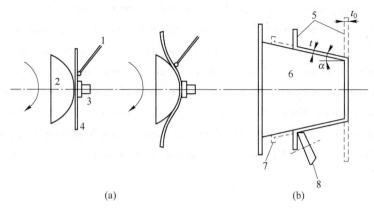

Fig. 10-5 Schematic representation of spinning processes
(a) manual spinning; (b) shear spinning
1—toll; 2—block; 3—clamp; 4, 5—blank; 6—mandrel; 7—cone; 8—roller

10.3 Shearing and Blanking

Shearing is the separation of metal by two blades moving as shown in Fig. 10-6(a). In shearing, a narrow strip of metal is severely plastically deformed to the point where it fractures at the surfaces in contact with the blades. The fracture then propagates inward to provide complete separation. The depth to which the punch must penetrate to produce complete shearing is directly related to the ductility of the metal. The penetration is only a small fraction of the sheet thickness for *brittle materials*, while for very *ductile materials* it may be slightly greater than the thickness.

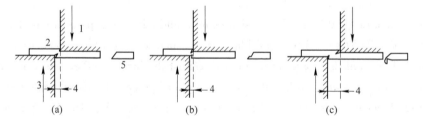

Fig. 10-6 Shearing of metal
(a) proper clearance; (b) insufficient clearance; (c) excessive clearance
1—moving blade; 2—metal; 3—stationary blade; 4—clearance; 5—*sheared edge*

The clearance between the blades is an important variable in shearing operations. With the proper clearance the cracks that initiate at the edges of the blades will propagate through the metal and meet near the center of the thickness to provide a clean *fracture surface* (Fig. 10-6(a)). Note that even with proper clearance there is still distortion at a sheared edge. Insufficient clearance will produce a *ragged fracture* (Fig. 10-6(b)) and also will require more energy to shear the metal than when there is proper clearance. With excessive clearance there is greater distortion of the edge, and more energy is required because more metal must plastically deform before it fractures. Furthermore, with too large a clearance, *burrs* or sharp projections are likely to form on the sheared edge. (Fig. 10-6(c)). A dull *cutting edge* also increases the tendency for the formation of burrs. The height of the burr increases with increasing clearance and increasing ductility of the metal. Because the quality of the sheared edge influences the formability of the part the control of clearance is important. Clearances generally range between 2% and 10% of the thickness of the sheet: the thicker the sheet the larger the clearance.

Neglecting friction, the force required to shear a metal sheet is the *product* of the length cut, the sheet thickness, and the *shearing strength* of the metal. Empirically, the maximum punch force to produce shearing is given by

$$p_{max} \approx 0.7 \sigma_b h L$$

where σ_b = the *ultimate tensile strength*;

h = sheet thickness;

L = total length of the sheared edge.

The shearing force can be reduced appreciably by making the edges of the cutting tool at an inclined angle so that only a short part of the total length of cut is made at one time. The *bevel* of the cutting edge is called shear.

A whole group of press operations are based on the process of shearing. The shearing of closed contours, when the metal inside the contour is the desired part, is called blanking. If the material inside the contour is discarded, then the operation is known as punching, or piercing. Punching *indentations* into the edge of the sheet is called *notching*. *Slitting* is a shearing cut which does not remove any metal from the sheet. Trimming is secondary operation in which previously formed parts at finished to size, usually by shearing excess metal around the periphery. The removal of forging flash in a press is a *trimming* operation. When the sheared edges of a part are trimmed or squared up by removing a thin shaving of metal the operation is called *shaving*.

Fine blanking is a process in which very smooth and *square edges* are produced in small parts such as gears, *cams*, and levers. To achieve this, the sheet metal is tightly locked in place to prevent distortion and is sheared with very small clearances on the order of 1% of the thickness at slow speeds. Usually the operation is carried out on a triple-action press so that the movements of the punches, hold down ring, and die can be controlled individually.

10.4 Bending

Bending is the process by which a straight length is transformed into a curved length. It is a very common forming process for changing sheet and plate into channel, *drums*, tanks, etc. In addition, bending is part of the deformation in many other forming operation. The definition of the terms used in bending are illustrated in Fig. 10-7. The bend radius R is defined as the radius of curvature on the concave, or inside, surface of the bend. For elastic bending below the elastic limit the strain passes through zero halfway through the thickness of the sheet at the *neutral axis*. In plastic bending beyond the elastic limit the neutral axis moves closer to the inside surface of the bend as the bending proceeds. Since the plastic strain is proportional to the distance from the neutral axis, fibers on the outer surface are strained more than fibers on the inner surface are contracted. A fiber at the mid-thickness is stretched, and since this is the average fiber, it follows that there must be a decrease in thickness (radial direction) at the bend to preserve the constancy of volume. The smaller the radius of curvature, the greater the decrease in thickness on bending.

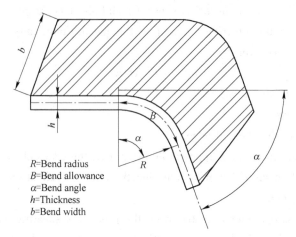

R=Bend radius
B=Bend allowance
α=Bend angle
h=Thickness
b=Bend width

Fig. 10-7 Definition of terms used in bending

10.5 Stretch Forming

Stretch forming is the process of forming by the application of primarily tensile forces in such a way as to stretch the material over a tool or form block. The process is an outgrowth of the *stretcher leveling* of rolled sheet. Stretch forming is used most extensively in the aircraft industry to produce parts of large radius of curvature, frequently with double curvature. An important consideration is that *springback* is largely eliminated in stretch forming because the *stress gradient* is relatively uniform. On the other hand, because tensile stresses predominate, large deformation can be obtained by this process only in materials with appreciable ductility.

Stretch-forming equipment consists basically of a hydraulically driven ram (usually vertical), which carries the punch or form block and two jaws for gripping the ends of the sheet(Fig. 10-8). No *female die* is used in stretch forming. The grips may be pivoted so that the tension force is always in line with the edge of the unsupported sheet, or they can be fixed, in which case a large radius is needed to prevent tearing the sheet at the jaws. In using a stretch-forming machine the sheet-metal blank is first bent or draped around the form block with relatively light tensile pull, the grips are applied, and the stretching load is increased until the blank is strained plastically to final shape. This differs from wrap forming (Sec. 10.2) since in the latter process the blank is first gripped and then while still straight is loaded to the *elastic limit* before wrapping around the form block.

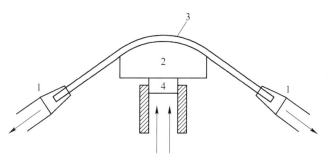

Fig. 10-8 Stretch-forming operation
1—jaw; 2—form block; 3—sheet; 4—ram

Stretching commonly is found as a part of many sheet-forming operations. For example, in forming a cup with a hemispherical bottom, the sheet is stretched over the punch face. Most complex automotive *stampings* involve a stretching component.

For a strip loaded in tension we have seen that the limit of uniform deformation occurs at a strain equal to the strain-hardening *exponent*. In *biaxial tension*, the necking, which occurs in biaxial tension, is inhibited if $\dfrac{\sigma_2}{\sigma_1} > \dfrac{1}{2}$, and instead the material develops *diffuse necking*, which is not highly localized or readily visible to the eye. Eventually in the stretching of a thin sheet plastic instability will occur in the form of a narrow localized neck. There will be a direction of zero-length increase inclined at an angle ϕ to the deforming axis (Fig. 10-9).

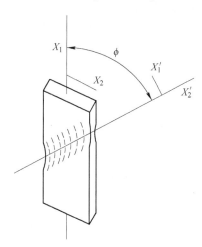

Fig. 10-9 Localized necking in a strip loaded in tension.

The *normal strain* ε'_2 must be zero, for it were not, material adjoining the edges of the band would have to deform and the band would spread out along X'_1 and the band would grow into a diffuse neck. It can be shown that $\phi \approx \pm 55°$ for an *isotropic material* in pure tension. Moreover, the criterion for local necking was shown in other Section to be $\dfrac{d\sigma}{d\varepsilon} = \dfrac{\sigma}{2}$, which arises from the fact that the area decreases with straining in local necking less rapidly than with diffuse necking. For *power-law* strain hardening, we have seen that for diffuse necking $\varepsilon_u = n$, but for local necking $\varepsilon_u = 2n$. The normal strain along X'_2 must be zero.

10.6 Deep Drawing

Deep drawing is the metalworking process used for shaping flat sheets into cup-shaped articles such as *bathtubs*, shell cases, and automobile panels. This is done by placing a blank of appropriate size over a shaped die and pressing the metal into the die with a punch (Fig. 10-10). Generally a clamping or hold-down pressure is required to press the blank against the die to prevent wrinkling. This is best done by means of a *blank holder* or hold-down ring in a double-action press. Although the factors which control the deep-drawing process are quite evident, they interact in such a complex way that precise mathematical description of the process is not possible in simple terms. The greatest amount of experimental and analytical work has been done on the deep drawing of a flat-bottom cylindrical cup from a flat circular blank. The discussion of deep drawing, which follows will be limited to this relatively simple situation.

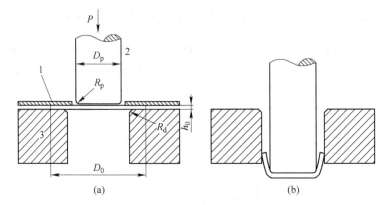

Fig. 10-10 Deep drawing of a cylindrical cup
(a) before drawing; (b) after drawing
1—holddown ring; 2—punch; 3—die

In the deep drawing of a cup the metal is subjected to three different types of deformations. The metal at the center of the blank under the head of the punch is wrapped around the profile on the punch, and in so doing it is thinned down. The metal in this region is subjected to biaxial tensile stress due to the action of the punch. Metal in the outer portion of the blank is drawn radically inward the throat of the die. As it is drawn in, the outer circumference must continuously decrease from that of the original blank πD_0 to that of the finished cup πD_p. This means that it is subjected to a compressive strain in the circumferential, or *hoop*, direction and a tensile strain in the radial direction. As a result of these two principal strains, there is continual increase in the thickness as the metal moves inward. However, as the metal passes over the die radius, it is first bent and then straightened while at the same time being subjected to a tensile stress. This plastic bending under tension results in considerable thinning, which modifies the thickening due to the circumferential shrinking. Between the inner stretched zone and the outer shrunk zone there is a narrow ring of metal which has not been bent over either the punch or the die. The metal in this region is subjected only to simple tensile loading throughout the drawing operation.

If the clearance between the punch and the die is less than the thickness produced by free thickening, the metal in these regions will be squeezed, or ironed, between the punch and the die to produce a uniform wall thickness. In commercial deep drawing clearances about 10% to 20% greater than the metal thickness are common. Ironing operations in which appreciable uniform reductions are made in the wall thickness use much smaller clearances.

The force on the punch required to produce a cup is the summation of the ideal force of deformation, the frictional forces, and the force required to produce ironing (if present). Fig. 10-11 illustrates the way in which these components of the total punch force way with the length of stroke of the punch. The ideal force of deformation increases continuously with length of travel because the strain is increasing and the flow stress is increasing owing to strain hardening. A major contribution to the friction force comes from the hold-down pressure. This force component peaks early and decreases with in creasing travel because the area of the blank under the hold-down ring is continually decreasing. Any force required to produce ironing occurs late in the process after the cup wall has reached maximum thickness. An additional factor is the force required to bend and unbend the metal around the radius of the die. One measurement of the work required in cupping showed that 70% of the work went into the radial drawing of the metal, 13% into overcoming friction, and 17% into the bending and unbending around the die radius.

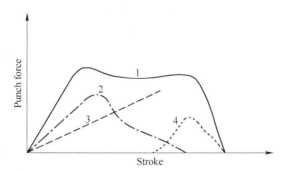

Fig. 10-11 Punch forces vs punch stroke for deep drawing
1—total punch force; 2—friction; 3—ideal deformation; 4—ironing

Words and Expressions

flat blank　扁坯，板坯
stretch flange　伸长类翻边，拉伸凸缘
shrink flange　收缩法兰，压缩类翻边
deep-recessed　深腔
shallow-recessed　浅腔

bead　筋，压条，车轮圆缘，珠状
emboss　压花，压纹
corrugate　折皱，波纹的
severity　严重强度，刚度
shearing　剪切，切变

deep drawing 深冲
ironing 变薄拉深，减径挤压
bulk-forming 体积成型
sheet forming 板料成型
tensile force 张力，拉力
compressive force 压力，挤压力，压缩力
buckle 翘曲
fold 折叠
wrinkle 起皱
necking 颈缩
failure 断裂
stretching 伸延，张延
finishing operation 精整
slide 滑块
downstroke 下行冲程
single-action press 单动压力机
double-action press 双动压力机
hold-down 压具，夹爪，压料
punch 冲头
subpress 半成品压力机，小压力机
die set 模组，冲模定位架，模架
clamping pressure 合模压力，夹紧压力
hold-down ring 压紧环，压边圈
air cylinder 汽缸
progressive forming 级进成型，连续成型
washer 垫圈
blank 落料，冲裁
piercing 冲孔
stripper plate 卸料板，脱模板
compound dies 复合模
transfer die 自动模
zinc-base alloy 锌基合金
Kirksite 锌合金
epoxy resin 环氧树脂
press brake 折弯机
roll forming 滚压成型，成型轧制
Rubber hydroforming 橡皮液压成型
form block 成型模，试印用压凸模
polyurethane 聚氨基甲酸酯

retainer 护圈，隔栅，护套
Guerin process 格林橡胶模冲压法，金属薄板成型法
rubber bag 橡皮包套
platen 压板
Rubber forming 橡皮成型
positioning pin 定位销
bending rolls 辊子弯板机，弯板机，弯曲辊，滚弯机
thin-gage 薄壁
curvature 曲率，弧度
bending moment 弯曲力矩
wiper-type 滑动，接触，瞬动
wiper roll 擦拭辊，滑动辊
hydraulic cylinder 液压缸
wrap forming 卷缠成型，拉伸成型
forming limit 成型极限
mandrel 心轴，芯棒
tank head 油罐端板
spinning 旋压
shear-spinning 剪切旋压，变薄旋压
power spinning 强力旋压
flow turning 变薄旋压
hydrospinning 液力旋压成型法
curvilinear 曲线
rocket motor casing 火箭发动机壳
missile nose cone 火箭或飞弹之鼻锥体，头锥体
tube spinning 管材旋压、筒表变薄旋压
explosive forming 爆炸成型
detonate 爆破，引爆
standoff distance 偏距，远距
sheared edge 剪切毛边，剪切边，剪切端面
explosive charge 炸药
brittle material 脆性材料
ductile material 塑性材料
ragged fracture 锯齿状断口，粗糙断口
fracture surface 断裂面，断面

burr　毛刺
cutting edge　刃口
product　乘积
shearing strength　抗剪强度
ultimate tensile strength
　　极限抗拉强度
bevel　斜面，削面
notching　切口，下凹
indentation　凹槽，压痕
slitting　纵切，纵剪，开缝
trimming　修边，剪切
periphery　周边，边界
shaving　修整，整修
square edge　直角边，直角边缘
cam　凸轮
fine blanking　精密冲裁
drum　鼓形，筒

neutral axis　中性轴
female die　阴模，凹模
stretcher leveling　拉伸矫直
springback　回弹，弹性后效
elastic limit　弹性极限
stampings　冲压，冲压件
exponent　指数
stress gradient　应力梯度
biaxial tension　双向拉伸
diffuse necking　扩散颈缩
normal strain　正应变，法向应变
isotropic material　各向同性材料
power-law　幂律
bathtub　浴缸，缸状物
blank holder
　　防皱压板，空套，坯夹，支架
hoop　环

11 Extrusion

11.1 Classification of Extrusion Processes

Extrusion is the process by which a block of metal is reduced in cross section by forcing it to flow through a *die orifice* under high pressure. In general, extrusion is used to produce cylindrical bars or hollow tubes, but shapes of irregular cross section may be produced from the more readily extrudable metals, like aluminum. Because of the large forces required in extrusion, most metals are extruded hot under conditions where the deformation resistance of the metal is low. However, cold extrusion is possible for many metals and has become an important commercial process. The reaction of the extrusion billet with the *container* and die results in high compressive stresses, which are effective in reducing the cracking of materials during *primary breakdown* from the ingot. This is an important reason for the increased utilization of extrusion in the working of metals difficult to form, like stainless steels, *nickel – based alloys*, and other high – temperature materials.

The two basic types of extrusion are direct extrusion and indirect extrusion (also called inverted, or back, extrusion). Fig. 11-1(a) illustrates the process of direct extrusion. The metal billet is placed in a container and driven through the die by the ram. A *dummy block*, or *pressure plate*, is placed at the end of the ram in contact with the billet. Fig. 11-1(b) illustrates the *indirect extrusion* process. A hollow ram carries the die, while the other end of the container is closed with a plate. Frequently, for indirect extrusion, the ram containing the die is kept stationary, and the container with the billet is caused to move. Because there is no relative motion between the wall

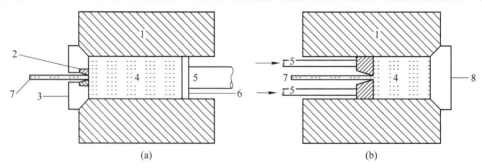

Fig. 11-1 Types of extrusion
(a) direct extrusion; (b) indirect extrusion

1—container; 2—die; 3—*die holder*; 4—billet; 5—ram; 6—dummy block; 7—extrusion; 8—*closure plate*

of the container and the billet in indirect extrusion, the friction forces are lower and the power required for extrusion is less than for direct extrusion. However, there are practical limitations to indirect extrusion because the requirement for using a hollow ram limits the loads which can be applied.

Tubes can be produced by extrusion by attaching a mandrel to the end of the ram. The clearance between the mandrel and the die wall determines the wall thickness of the tube. Tubes are produced either by starting with a hollow billet or by a two-step extrusion operation in which a solid billet is first pierced and then extruded.

Extrusion was originally applied to the making of lead pipe and later to the *lead sheathing* of cable. Fig. 11-2 illustrates the extrusion of a lead sheath on *electrical cables*.

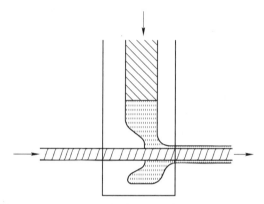

Fig. 11-2 Extrusion of a lead sheath on electrical cable

Impact extrusion is a process used to produce short lengths of hollow shapes, such as collapsible toothpaste tubes. It may be either indirect or direct extrusion, and it is usually performed on a light-speed mechanical press. Although the process generally is performed cold, considerable heating result from the high-speed deformation. Impact extrusion is restricted to the softer metals such as lead, *tin*, aluminum, and copper.

11.2 Extrusion Equipment

Most extrusions are made with hydraulic presses. Hydraulic extrusion presses are classified into horizontal and vertical presses, depending upon the direction of travel of the ram. Vertical extrusion presses are generally built with capacities of 300 tons to 2000 tons. They have the advantages of easier alignment between the press ram and the tools, higher rate of production, and the need for less floor space than horizontal presses. However, they need considerable *headroom*, and to make extrusions of appreciable length, a floor pit is frequently necessary. Vertical presses will produce uniform cooling of the billet in the container, and thus symmetrically uniform deformation will result. In a horizontal extrusion press the bottom of the billet which lies in contact with the container will cool more rapidly than the top surface, unless the extrusion container is internally heated, and therefore the deformation will be nonuniform. Warping of bars will result, and nonuniform wall thickness will occur in tubes. In commercial operations the chief use for vertical presses is in the production of thin-wall tubing, where uniform wall thickness and *concentricity* are required. Horizontal extrusion presses are used for most commercial extrusion of bars and shapes. Presses with a capacity of 1500 tons to 5000 tons are in regular operation, while a few presses of 14000 tons capacity have been constructed.

The ram speed of the press can be an important consideration since high ram speeds are required in high-temperature extrusion where there is a problem of heat transfer from the billet to the tools. Ram speeds of 254m/min to 381m/min may be used in extruding *refractory metals*, and this requires a *hydraulic accumulator* system with the press. At the other extreme, aluminum and copper alloys are prone to hot shortness so that the ram speed must be restricted to a few inches per minute. In this case, direct-drive pumping systems are adequate. Work has been done on presses equipped for preselected program control of ram speed in order to maintain a uniform finishing temperature.

The dies and tooling used in extrusion must withstand considerable abuse from the high stresses, *thermal shock*, and *oxidation*. Fig. 11-3 illustrates a typical extrusion tooling assembly. This assembly is designed for easy replacement of damaged parts and for reworking and reuse of components of the tooling. The die stack consists of the die (6), made from highly alloyed tool steel, which is supported by a die holder (5) and a *bolster* (7), all of which are held in a die head (2). This entire assembly is sealed against the container on a *conical seating surface* by the pressure applied by a wedge (1). Since the extrusion container must withstand high pressures, it is usually made in two parts. A liner (4) is shrunk into the more massive container (3) to produce a compressive *prestress* in the inside surface of the liner. The *extrusion ram* or stem is usually highly loaded in compression. It is protected from the hot billet by a *follower pad* immediately behind the billet. Since the liner and follower pad are subjected to many cycles of thermal shock, they will need to be replaced periodically.

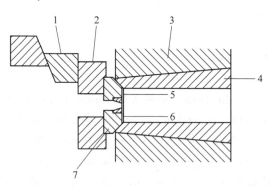

Fig. 11-3 Typical arrangement of extrusion tooling
1—wedge; 2—*die head*; 3—container; 4—*liner*; 5—die holder; 6—die; 7—bolster

There are two general types of extrusion dies. *Flat-faced dies* (Fig. 11-4(a)) are used when the metal entering the die forms a *dead zone* and shears internally to form its own *die angle*. A parallel land on the exit side of the die helps strengthen the die and allows for reworking of the flat face on the entrance side of the die without increasing the exit diameter. Dies with conical *entrance angles* (Fig. 11-4(b)) are used in extrusion with good lubrication. Decreasing the die angle increases the *homogeneity* of deformation and lowers the extrusion pressure, but beyond a point the friction in the die surfaces becomes too great. For most extrusion operations the optimum semidie angle (α) is between 45° and 60°.

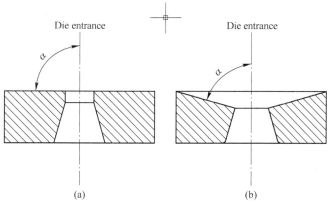

Fig. 11-4　Typical extrusion dies
(a) flat-faced (square) die; (b) *conical die*

In addition to the extrusion press, billet-heating facilities are needed, and for production operations, there should be automatic transfer equipment for placing the heated billet in the container. Provision for heating the extrusion container may also be required. A *hot saw* is needed to cut off the extrusion so that the discard, or *butt*, can be removed from the die. Alternatively, extrusion may be done with a *carbon block* between the billet and the *follower*. This allows complete extrusion of the billet without a butt. Finally, there must be a *runout table* for catching the extrusion and a *straightener* to correct minor *warpage* in the extruded product.

11.3　Hot Extrusion

The principal variables which influence the force required to cause extrusion are (1) the type of extrusion (direct vs. indirect), (2) the *extrusion ratio*. (3) The working temperature, (4) the speed of deformation, and (5) the frictional conditions at the die and container wall.

In Fig. 11-5 the extrusion pressure is plotted against ram travel for direct and indirect *intrusion*. Extrusion pressure is the extrusion force divided by the cross-sectional area of the billet. The rapid rise in pressure during the initial ram travel is due to the initial compression of the billet to fill the extrusion container. For direct extrusion the metal begins to flow through the die at the maximum value of pressure, the *breakthrough* pressure. As the billet extrudes through the die the pressure required to maintain flow progressively decreases with decreasing length of the billet in the container. For indirect extrusion there is no relative motion between the billet and the container wall. Therefore, the extrusion pressure is approximately constant with increasing ram travel and represents the stress required to deform the metal through the die. While this appears to be an attractive process, in practice it is limited by the need to use a hollow ram which creates limitations on the size of the extrusion and the extrusion pressures which can be achieved. Therefore, most hot extrusion is done by the direct process. Finally, returning to Fig. 11-5, at the end of the stroke the pressure builds up rapidly and it is usual to stop the ram travel so as to leave a small discard in the container. This discard often contains defects, which are unwanted in the product.

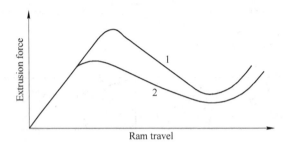

Fig. 11-5 Typical curve of extrusion pressure vs. ram travel for direct and indirect extrusion
1—direction extrusion; 2—indirect extrusion

The extrusion ratio is the ratio of the initial cross-sectional area of the billet to the final cross-sectional area after extrusion. $R = A_0/A_f$. Extrusion ratios reach about 40 : 1 for hot extrusion of steel and may be as high as 400 : 1 for aluminum. It is important to appreciate the distinction between the fractional reduction in area., $r = 1 - A_f/A_0$, and the extrusion ratio R, $R = 1/(1-r)$. For large deformations R is a more descriptive parameter. For example, the change in fractional reduction from 0.95 to 0.98 appears relatively minor, yet it corresponds to a change in area ratio from $R = 20 : 1$ to $R = 50 : 1$.

Because there is constancy of *mass flow rate* through the die, the velocity of the extruded product is the ram velocity $\times R$, so that quite high sliding velocities can be achieved along the die land. The extrusion pressure is directly related to the natural logarithm of the extrusion ratio, so that the extrusion force may be expressed as

$$P = kA_0 \ln A_0/A_f \qquad (11-1)$$

where $k=$ the "extrusion constant" an overall factor which accounts for the flow stress, friction, and inhomogeneous deformation.

Most metals are extruded hot so as to take advantage of the decrease in flow stress or deformation resistance with increasing temperature. Since hot-working introduces the problems of oxidation of the billet and the extrusion tools and softening of the die and tools, as well as making it more difficult to provide adequate lubrication, it is advantageous to use the minimum temperature which will provide the metal with suitable plasticity. The upper hot-working temperature is the temperature at which hot shortness occurs, or, for pure metals, the *melting point*. Because of the extensive deformation produced in extrusion, considerable internal heating of the metal also results. Therefore, the top working temperature should be safely below the melting-point or hot-shortness range.

In the extrusion of steel the billets are heated in the range 1100℃ to 1200℃, which the tooling is preheated to around 350℃. The extrusion pressures generally are in the range of 861.9MPa to 1241.1MPa. The combination of high stresses and temperature place a severe demand on the *glass lubrication* system. The hot extrusion of lower-strength aluminum alloys is accomplished without any billet lubrication. The billet shears within itself near the container wall to create its own internal conical die surface. Metal deformation is very nonuniform, which results in a wide variation in *heat-treatment* response.

The interrelationship of temperature, deformation velocity, and deformation (reduction) to define conditions for successful crack-free hot-working was considered in other Section and illustrated with other Figure. This situation is very typical of hot extrusion.

Increasing the ram speed produces an increase in the extrusion pressure. A tenfold increase in the speed results in about a 50% increase in pressure. Greater cooling of the billet occurs at low extrusion speeds. When this becomes pronounced, the pressure required for direct extrusion will actually increase with increasing ram travel because of the increased flow stress as the billet cools. The higher the temperature of the billet, the greater the effect of low extrusion speed on the cooling of the billet. Therefore, high extrusion speeds are required with high-strength alloys that need high extrusion temperatures. The temperature rise due to deformation of the metal is greater at high extrusion speeds, and therefore problems with hot shortness may be accentuated.

The selection of the proper extrusion speed and temperature is best determined by *trial and error* for each alloy and billet size. The interdependence of these factors is shown schematically in other Figure. For a given extrusion pressure, the extrusion ratio increases with increasing temperature. For any given temperature a large extrusion ratio can be obtained with a higher pressure. The maximum billet temperature, on the assumption that there are no limitations from the strength of the tools and die, is determined by the temperature at which *incipient melting* or hot shortness occurs in the extrusion. The temperature rise of the extrusions will be determined by the speed of extrusion and the amount of deformation (extrusion ratio). Therefore, the curve which represents the upper limit to the safe extrusion at infinite speed, where none of the heat produced by deformation is dissipated. At lower extrusion speeds there is greater heat dissipation, and the allowable extrusion ratio for a given preheat temperature increase. The allowable extrusion range is the region under the curve of constant pressure and extrusion speed.

11.4 Cold Extrusion and Cold Forming

Cold extrusion is concerned with the cold forming rod and bar stock of small machine parts, such as *spark plug bodies*, shafts, pins, and hollow cylinders or cans. More properly, the subject should be *expanded* to include other cold-forming processes (such as upsetting, expanding, and *coining*) than are not strictly extrusion. Ax symmetric parts are particularly suited to cold-forming processes. Precision cold-forming can result in high production of parts with good dimensional control and good surface finish. Because of extensive strain hardening, it often is possible to use cheaper materials with lower alloy content. Extensive use is made of cold-formed low-alloy steels in the automotive industry.

The first major application of cold-forming of steel was the development during Word War II of a process for making steel *cartridge cases*. The key was the development of a suitable lubricant system. For steel, a *zinc phosphate conversion coating* and soap is usually preferred. Other important factors are the use of steel with high resistance to *ductile fracture* and the design of the tooling to minimize tensile-stress concentrations. The extensive literature on cold-forming has been reviewed by Watkins.

11.5 Hydrostatic Extrusion

The concept of *hydrostatic extrusion* was introduced in other figure. Because the billet is subjected to uniform *hydrostatic* pressure, it does not upset to fill the *bore* of the container as it would in conventional extrusion. This means that the billet may have a large length-to-diameter ratio (even coils of wire can be extruded) or it may have an irregular cross section. Since there is no container billet friction, the curve of the extrusion pressure vs. ram travel is nearly flat, like that for indirect extrusion in Fig. 11-5. Because of the pressurized fluid, lubrication is very effective, and the extruded product has good surface finish and dimensional accuracy. Since friction is nearly absent, it is possible to use dies with a very low *semicone angle* ($\alpha = 20°$), which greatly minimizes the redundant deformation.

Because hydrostatic extrusion employs a pressurized fluid, there is an inherent limitation to hot-working with this process. A practical limit on fluid pressure of around 1723.8MPa currently exists because of the strength of the container and the requirement that the fluid not solidify at high pressure. This limits the obtainable extrusion ratio for mild steel to less than 20 : 1, while for very soft metals like aluminum, it is possible to achieve extrusion ratios in excess of 200 : 1. The extrusion of wire is one of the areas actively being pursued. Because of the large amount of stored energy in a pressurized fluid, the control of the extrusion on the exit from the die may be a problem. However, this is solved by augmented hydrostatic extrusion in which an axial force is applied either to the billet or to the extrusion. The fluid pressure is kept at less than the value required to cause extrusion, and the balance is provided by the augmenting force. In this way, much better control is obtained over the movement of the extrusion.

A number of methods have been developed to increase the rate of production of hydrostatic extrusion or to place it on a continuous basis. *Thick-film* hydrostatic extrusion minimizes the amount of pressurized fluid, and because the billets can be *precoated* with the hydrostatic medium, the production rate approaches that for conventional extrusion. Fuchs developed a continuous extrusion process which uses *viscous drag* of a flowing polymer to feed the billet rod into an *extrusion chamber* and through the extrusion die. Problems with pressure control have led to a modified design of a continuous *gear-driven extruder*. Alexander has discussed continuous hydrostatic extrusion from a general point of view.

11.6 Extrusion of Tubing

With modern equipment, tubing may be produced by extrusion to tolerance as close as those obtained by *cold-drawing*. To produce tubing by extrusion; a mandrel must be fastened to end of the extrusion ram. The mandrel extends to the entrance of the extrusion die, and the clearance between the mandrel and the die wall determines the wall thickness of the extruded tube. Generally, a hollow billet must be used to so that the mandrel can extend to the die. In order to produce concentric tubes, the ram and mandrel must move in axial alignment with the container and the die. Also, the axial hole in the billet must be concentric, and the billet should offer equal resistance to deformation over its cross section.

One method of extruding a tube is to use a hollow billet for the starting material. The hole may be produced either by casting, by machining, or by hot piercing in a separate press. Since the bore of the hole will become oxidized during heating, the use of a hollow billet may result in a tube with an oxidized inside surface.

A more satisfactory method of extruding a tube is to use a solid billet, which is pierced and extruded in one step in the extrusion press. With a modern extrusion press the piercing mandrel is actuated by a separate hydraulic system from the one which operates the ram. The piercing mandrel moves coaxially with the ram, but it is independent of its motion (Fig. 11-6). In the operation of a double-action extrusion press the first step is to upset the billet with the ram while the piercing mandrel is withdrawn. Next the billet is pierced with the pointed mandrel, ejecting a metal plug through the die. Then the ram advances and extrudes the billet over the mandrel to produce a tube.

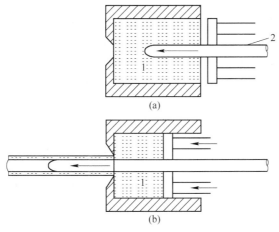

Fig. 11-6 Tube extrusion
(a) piercing; (b) extrusion
1—billet; 2—piercing mandrel

A third method of extruding tubing, which is used with aluminum and *magnesium alloys*, is to use a solid billet and a porthole die with a standard extrusion ram without a mandrel. A sketch of a *porthole die* is shown in Fig. 11-7. The metal is forced to flow into separate streams and around the central bridge, which supports a short mandrel. The separate streams of metal which flow through the ports are brought together in a welding chamber surrounding the mandrel, and the metal exits from the die as a tube. Because the separate metal streams are joined within the die, where there is no *atmospheric contamination*, a perfectly sound weld is obtained. In addition to tubing porthole extrusion is used to produce hollow unsymmetrical shapes in aluminum alloys.

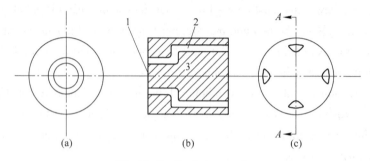

Fig. 11-7 *Porthole extrusion*
(a) exit face; (b) cross section A—A; (c) entrance face
1—mandrel; 2—welding chamber; 3—bridge

Words and Expressions

die orifice　模口
container　挤压筒
primary breakdown
　一次击穿，原始组织破坏
nickel-based alloy　镍基合金
die holder　模座，模架，凹模固定板
dummy block　挤压垫
closure plate　堵板，闭合板
direct extrusion　正挤压，直接挤压
pressure plate　压板
indirect extrusion　反挤压，间接挤压
lead sheath　铅护套
electrical cable　电缆
tin　锡
headroom　净空，头上空间
impact extrusion　冲击挤压，冲挤
concentricity　同轴度，同心度
refractory metal　耐高温金属

hydraulic accumulator　液压蓄能器
thermal shock　热冲击
oxidation　氧化作用
bolster　垫块
conical seating surface　锥形支持面
die head　模头
liner　衬圈，衬套，垫圈
extrusion ram
　挤压杆，挤压冲头，挤压凸模
prestress　预压力
follower pad　挤压垫片
Flat-faced die　平模
dead zone　死区，盲区
die angle　模口角度，模切角，模孔锥角
conical die　拉模，锥孔模
homogeneity　均匀性
entrance angle　入口锥
hot saw　热锯

butt　坯，锭
follower　跟随器，随动机构
carbon block　炭精块，大块炭砖
straightener　矫直器，矫直装置
warpage　翘曲
runout table　输出滚道
intrusion　侵入，干涉
breakthrough　临界，穿透
extrusion ratio　挤压比
mass flow rate　质量流率，质量流速
melting point　熔点
glass lubrication　玻璃润滑
heat-treatment　热处理
trial and error　累试法，反复实验法
incipient melting　初熔
spark plug body　火花塞壳
expand　胀形
coining　精压，冲制，印压
cartridge case　弹夹

conversion coating　转化涂层，化学覆层
zinc phosphate　磷酸锌
ductile fracture　塑性破坏，韧性断裂
Hydrostatic Extrusion　静液力挤压
hydrostatic　静水
bore　腔
semicone angle　计锥角
mild steel　低碳钢，软钢
thick-film　厚膜
precoat　预涂
viscous drag　黏性阻力
extrusion chamber　挤压筒
gear-driven extruder　齿轮传动的挤压机
cold-drawing　冷拉
Porthole extrusion　多孔分流模挤压
magnesium alloy　镁合金
porthole die　异型孔挤压模
atmospheric contamination　空气污染

12 Classification of Welding Processes

Welding processes may be classified according to the way in which the four basic requirements—particularly the first three—are satisfied. The energy for welding is almost always supplied as heat so that divisions can be made according to the methods by which the heat is generated locally. These methods may be defined and grouped as follows:

(1) *Mechanical.* Heat generated by impact or friction or liberated by the elastic or plastic deformation of the metal.

(2) *Thermochemical. Exothermic reactions*, *flames* and *arc plasmas.* It is necessary to explain why plasmas should be put in the same class as *oxyfuel gas flames.* Although chemical reactions may not take place in a plasma the method of heat transfer to the work is the same as for processes employing an envelope of burning gas. This holds for all processes in which the work does not form part of the *arc circuit.* The so-called *nontransferred arc* produces a plasma flame. Whereas the *transferred arc* is a *constricted arc* and falls in the arc-process category.

(3) *Electric arc.* Both A. C. and D. C. arcs with electrodes which melt and those which do not.

(4) *Radiant energy.* This category is suggested to cover the new processes such as *laser* and *electron beam welding* and others which may yet be developed. The essential feature of a radiant energy process is that energy is focused on the workpiece and heat is generated only where the focused beam is intercepted.

It is not possible to define all welding processes completely by the source of thermal energy. This applies particularly to the many variations of arc welding and it is customary to complete the definition by reference to the way the process satisfies the condition of atmosphere control. All welding processes can be examined in the same way by placing the names of the processes within a grid formed by listing the sources of heat along one axis, as is done in Table 12-1. The diagram can now be divided up into areas enclosing processes with a basic similarity. Seven such areas are readily identified corresponding to processes as follows: (1) solid phase, (2) thermochemical, (3) electric resistance, (4) unshielded arc, (5) flux shielded arc, (6) gas shielded arc, (7) radiant energy.

Certain areas in the diagram can be marked out as regions where welding processes could not exist—for example flames cannot be used in vacuum.

This way of classifying welding processes is less rigid than the family-tree method and makes it possible to account for certain anomalies. The resistance butt-welding process, for example, while truly a solid-phase welding process, is normally included in the resistance-welding category. In the Table 12-1, the position of this process is clarified by drawing the boundary of the group (1) solid phase processes to include resistance butt and to exclude the remaining

12 Classification of Welding Processes

Table 12-1 Grouping of welding processes according to heat source and shielding method

Source of heat		Welding process classification						
		Shielding method					
		Vacuum	Inert gas	Gas	Flux	No shieding	Mechanica exclusion
No heat or heat by conduction		Cold pressure	Thermo-Compression bonding				Hot pressure Cold pressure
Mechanical		Explosive		(1)		Explosive	Friction Ultrasonic
Thermo-chemical	Flames plasma		Plasma	Atomic Hydrogen	Gas	Forge	Pressure butt
	Exothermic reactions			(2)	Thermit		
Electric resistance	Induction				(3)	H. F. induction	Induction butt
	Direct				Electro-slag	Flash butt H. f. Resistance projection	Spot Seam Or resistance butt
Electric arc	Consumable electrode		Inert-gas Metal-arc	CO_2 metal Arc Gas/flux Metal-arc	Covered Electrode Submerged-arc	Bare wire Stud Spark Discharge Percussion	
	Non consumable		Inert-gas Tungsten arc	(6)	(5)	(4) Carbon arc	
Radiant energy	Electro Magnetic			(7)		Laser	
	Particle	Electron-beam					

resistance processes. Similarly, electric slag welding and its derivatives can be placed correctly in the resistance heat source grid, but may be linked with the flux shielded arc processes with which they have a great deal in common.

There is no uniform method of naming welding processes. Many processes are named according to the heat source or shielding method, but certain specialized processes are named after the type of joint produced. Examples are *stud*, spot and butt welding. An overall classification can not take account of this because the same type of joint may be produced by a variety of processes. Stud welding may be done by arc or projection welding and spot welding by electric resistance, arc, or electron beam processes. But welding may be done by resistance, flash or any of a number of other methods. Although in common usage many processes have abbreviated names, the full names often follow the pattern: first, a statement of the type of shielding (where mentioned); secondly, the type of heat or energy source; thirdly, the type of joint (where this is of specific and not general importance), see Table 12-2.

Table 12-2 The full names in common usage

Inert-gas	Tungsten-arc	Spot
(Unshielded)	Arc	Stud
	Resistance	Butt
	(Resistance)	Projection
(Vacuum)	Electron-beam	
(Flux-covered electrode)	Metal-arc	
	Friction	(Butt)

It is often necessary when referring to processes to mention the way they are used, particularly whether the operation is manual or automatic. The practical operation of welding can be divided into three main parts:

(1) The control of welding conditions, particularly arc length and electrode or *filler* wire feed rate and time.

(2) The movement and guiding of the electrode, *torch* or welding head along the weld line.

(3) The transfer or presentation of parts for welding.

Processes are described as manual, semi-automatic, depending on the extent to which the parts mentioned above are performed manually. Manual welding is understood to be that in which the welding variables are continuously controlled by the operator and the means for welding are held in the operator's hand. Semi-automatic welding is that in which there is automatic control of welding conditions such as arc length, rate of filler-wire addition and weld time, but the movement and guiding of the electrode, torch or welding head is done by hand. With automatic welding at least parts (1) and (2) of the operation must be done by the machine. As *feedback control* devices are introduced and welding takes its place more frequently in the automatic production line, other definitions will be required.

Words and Expressions

mechanical   机械黏附法
thermochemical   热化学反应法
exothermic reactions   放热反应法
flames   火焰加热法
arc plasmas   电弧等离子加热法
oxyfuel gas flames   乙炔气体火焰
arc circuit   电弧电路
nontransferred arc
   非转移电弧，非过渡电弧

transferred arc   过渡电弧
constricted arc   压缩电弧
electric arc   电弧
radiant energy   辐射能
laser welding   激光焊接
electron beam welding   电子束焊接
thermit   铝热剂
electro-slag   电渣
spot welding   点焊

seam welding 缝焊
resistance welding 电阻焊
butt welding 对焊
stud welding 螺栓焊

filler 金属芯，填充物
torch 焊枪
feedback control
　　反馈控制，反馈调节

13 Methods of Welding

13.1 Types of Welded Joint

It is often extremely difficult to give accurate but brief definitions of common everyday things. This is very true of welding- daily we use equipments which rely on welded joints for there satisfactory operation, but rarely do we stop to consider the essential characteristics of the welds themselves. If we did, it would immediately become apparent that there are very many types of weld, which seem to vary according to the type of joint used. However, closer investigation would reveal that both the welds and the joints can be categorized into groups. We would discover that there are four basic types of joint: butt, 'T', *corner*, and *lap* (Fig. 13-1).

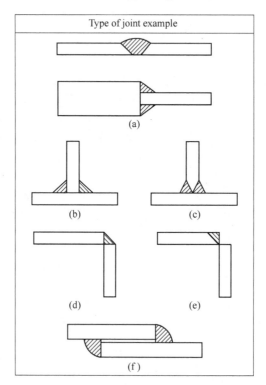

Fig. 13-1 Example of *welded joints*
(a) butt (unequal thickness fillet welded); (b) "T" (fillet welded); (c) "T" (butt welded)
(d) corner (fillet welded); (e) corner (butt welded); (f) lap (fillet welded)

The butt joint is characterized by the fact that the edges of the components are abutted and the load is transmitted along the common axis. This joint is of particular importance in fabrication, being used, for example, to join lengths of pipe, plates in ships' hulls, and *flanges* on bridge *girders* (Fig. 13-2).

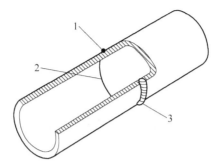

Fig. 13-2 Butt weld between lengths of pipe (sectioned to show bore)

1—weld run fused through the pipe wall; 2—weld penetrating into the bore; 3—surface of weld run raised above pipe surface

The "T" joint is probably the most commonly used connection in fabrication. Typical examples are at the flange to web junction in a plate girder, stiffeners welded to a panel, branches attached to a main pipe, and lifting lugs (Fig. 13-3). The joint can be made either with no penetration along the joint line (using "fillets" of weld metal to provide the load carrying connection) or with bonding across the interface, i.e. butt welded.

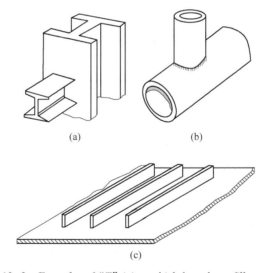

Fig. 13-3 Examples of "T" joints which have been fillet welded

(a) beam to column connection; (b) branch pipe welded to main pipe; (c) stiffeners welded to a panel

The corner joint can similarly be butt or fillet welded according to service requirements. Corner joints are normally associated with box sections, but it is worth noting that, from the point of view of both design and production, flange to pipe connections represent very important examples of corner joints.

Finally, the lap offers an interesting variant because it can be used in butt, "T", and corner joints. Most commonly, the bonding in a lap joint is over only a small area of the interface, either as a number of spots (Fig. 13-4) or as a narrow strip along the length of the joint. Lap joints are mainly used in sheet fabrication, and common items such as cars, washing machines, refrigerators, etc. contain many examples.

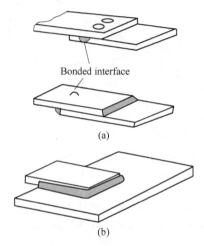

Fig. 13-4 Examples of lap joints
(a) spot welded; (b) fillet welded

An important feature of all the joints discussed above is the need for a bond which will transmit loads. This means that between the two components there must be a continuous metallic connection, i. e. a metallic bridge—which will be strong enough to withstand the stresses applied in service.

13.2 *Weld Formation*

We are now in a position to consider the definition of a weld, and in particular to look at the definition given in British Standard 499 : part 1 : 1965, "welding, brazing and thermal cutting glossary":

"A weld is a union between pieces of metal at faces rendered plastic or liquid by heat or by pressure or by both".

The first part of this definition follows from our discussion above: the union is the continuous load bearing metallic bridge to which we have referred. But what is the significance of the reference to rendering the faces plastic (i. e. permanently deformed) or liquid? To understand this we must first consider how the union is achieved.

Sufficient strength can be produced in a welded joint only by *interatomic bonding*, and the prime function of the welding operation is, therefore, to provide links between atoms at the interface of the joint. For these links to be formed, two conditions must be satisfied. Firstly, the surfaces must be in intimate contact. This implies that they should be atomically flat so that when they are brought together the gap between the respective surface atoms will be about equal to the atomic spacing within the metal, i.e. 124×10^{-10} m. Even polishing to a high finish with diamond dust is unlikely to produce this degree of flatness. Secondly, the surfaces must be metallurgically cleaned from any molecules of grease, paint, moisture, oxygen, or nitrogen present on practical welding system is to be based on the idea of simple interatomic bonding, a means of bringing the surfaces into intimate contact and, at the same time, dispersing the surface contaminants is essential.

13.3 *Cold Pressure Welding*

One way of achieving intimate contact and dispersal of contaminants is to force the surfaces together (Fig. 13-5). Under pressure the surfaces deform, breaking up the contaminants and bringing areas of clean metal into intimate contact.

The surface will prevent the metal atoms from uniting, even if intimate contact is achieved.

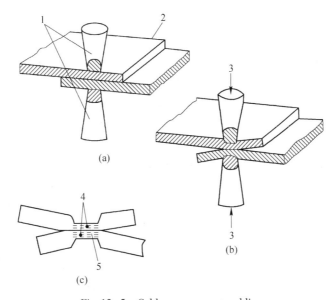

Fig. 13-5 Cold pressure spot welding
(a) joint assembly; (b) deformation produced by applying pressure; (c) cross section of weld
1—indentors; 2—sheets overlapped; 3—force; 4—regions of high deformation; 5—interface

The pressure required to disperse the contaminants leads to appreciable reductions in the thickness of the workpieces. Optimum joint strength is obtained at a level known as the threshold deformation, the actual value of which depends on the metals being welded : in general, the softer the metal, the lower the deformation required to initiate welding at room temperature.

Cold pressure welding is used to a limited extent to make welds between aluminium cable and connectors and for specialized applications such as welding caps to tubes (Fig. 13-6), but it is usually difficult to accommodate the amount of deformation required for the welding of commercial alloys.

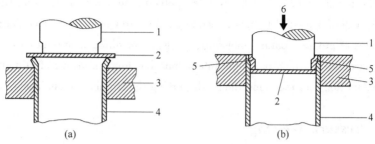

Fig. 13-6 Fitting caps to tubes by cold pressure welding
(a) joint assembly; (b) completed welding operation
1—ram; 2—cap; 3—die; 4—tube; 5—weld; 6—force

13.4 Hot Pressure Welding

Greater success can be achieved if the metals are heated during the welding operation. Raising the temperature reduces the value of the threshold deformation, and a number of successful hot pressure welding techniques have developed. Probably the oldest of these is forge welding, which has been in use by blacksmiths since about the year 1400 BC. In this process, the wrought iron or steel bars which are to be joined are heated to about 1350℃. At this temperature the iron oxides on the surface are melted and, when the components are hammered together, the molten oxides are squeezed out of the joint. Bonding then occurs at relatively low deformation levels.

Forge welding is now mainly used for craft work, but the principle of heating the component to make it easier to achieve a weld is also used in a number of modern developments of hot pressure welding. The method of heating is not critical, since the prime object is to raise the interfaces to a temperature at which the threshold value falls to about 25%. Since the hot metal is more plastic than it would be in cold pressure welds, the force which is required is smaller and it is possible to produce welds in hard metals such as steel. The resulting joint shows a similar structure to that of the cold pressure weld, but there is no evidence of cold working, since the deformation takes place above the *recrystallisation* temperature.

Of the various methods used to heat a joint for hot pressure welding, three of the most successful are gas heating, resistance heating, and induction heating.

13.5 Gas Heating

A ring burner is arranged around the joint, and mixed oxygen and *acetylene gas* is supplied to burners spaced uniformly around the circumference (Fig. 13-7).

13.6 Resistance Heating

This relies on the heating effect of a current flowing through a resistance (Fig. 13-8). For pre-

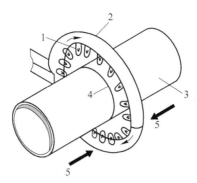

Fig. 13-7 Oxy acetylene pressure welding of small bore pipes
1—oxy acetylene; 2—ring burner; 3—pipe; 4—joint line; 5—force applied by clamps

ssure welding, the components, which are usually round bars, are held in clamps and a high current is passed along the work through the interface. Although the abutting ends of the bars are held in contact under pressure, the oxides at the interface offer a resistance to the passage of the welding current and heat is generated. The rate of heating depends on both the current and the resistance, and the latter is to some extent influenced by the magnitude of the end pressure which brings the faces into close contact, thus increasing the area which can conduct the electricity.

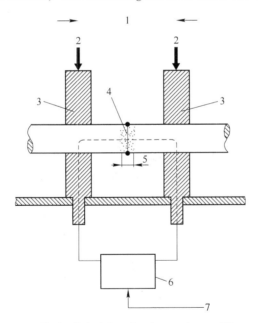

Fig. 13-8 Principles of resistance butt welding
1—clamps moved towards each other; 2—clamping force; 3—clamp; 4—interface heated by flow of current;
5—resistance heated zone; 6—transformer; 7—mains supply

13.7 *Induction Heating*

This involves the passage of a high-frequency current through a coil surrounding the components to be welded (Fig. 13-9). The magnetic field produced in the coil induces eddy currents in the steel workpieces, causing them to heat up.

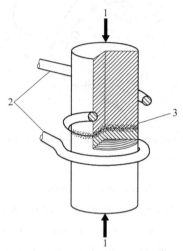

Fig. 13-9 Using a high frequency current to heat the interface in pressure welding
1—force; 2—coil carrying high frequency current; 3—joint area heated by induced eddy currents

Hot pressure welding is by no means a universally applicable technique, especially as it usually restricted to bars, rods, pipes, and narrow strips. It is also difficult to use for one-off work, since a number of trials must be made to ascertain the optimum combination of temperature and pressure.

13.8 Friction Welding

Without doubt, the most successful development in the field of pressure welding has been that of *friction welding*. The machine used for this process looks somewhat like a large lathe fitted with two chucks—one driven by a motor, the other fixed. The two parts to be joined are clamped in the chucks and one part is rotated (Fig. 13-10). This rotated component must be round in cross-

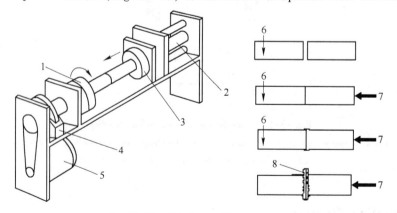

Fig. 13-10 Friction welding sequence
1—rotating chuck; 2—thrust cylinder; 3—sliding chuck; 4—brake; 5—motor; 6—rotation; 7—force; 8—flash

section, but the part held in the fixed chuck can be either a matching section or flat.

When the rotating chuck reaches the welding speed, the parts are brought into contact under a light axial load. As the abutting faces rub together, friction between them generates heat and localized hot plastic zones are produced. With the end load maintained, heat continues to be generated until the whole interface has reached a uniform temperature. At the same time, the plastic metal starts to flow outwards towards the periphery, carrying with it any oxides present at the joint face. When sufficient heating has occurred, the relative rotation of the parts is stopped rapidly and the end load may be increased. The result is a forged pressure butt weld having an excess metal flash which may be removed by machining. Weld times are short, being of the order of 20s to 100s.

The operating parameters for friction welding must be determined by trial and established before production welding begins, so it is clear the process is not ideally suited to one off fabrications. Where there is a repetitive element, however, it offers considerable potential and its attraction to the producer of engineering components is increased by the fact that friction welding units can be readily incorporated into workshops alongside lathes, milling machines, and similar machine tools. A good example of the use of friction welding in a high volume production line is the manufacture of axle casings for cars and heavy vehicles (Fig. 13-11).

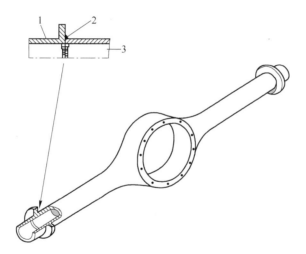

Fig. 13-11 Friction welding hub end forging to the axle casing for a heavy
1—hot forging; 2—friction weld (flash left in place); 3—axle casing

The main drawback in applying this process is that at least one of the components being joined must be round in cross section. The need to rotate one member of the joint can also pose problems. When welding long lengths of pipe, for example. In this case, a solution seems to lie in rotating a wedge-shaped insert between the abutting faces of the pipes, which are being forced towards each other. In this variation, known as radial friction welding, heat is generated at two interfaces and plastic deformation occurs as in the conventional process.

Words and Expressions

corner　角接
lap　搭接
welded joints　焊接接头
flange　法兰
girder　大梁，桁架
weld formation　焊接成型
interatomic bond　原子键力
cold pressure welding　冷压焊

hot pressure welding　热压焊
recrystallization　再结晶
gas heating　气焊
acetylene gas　乙炔气体
resistance heating　电阻加热
induction heating　感应加热
friction welding　摩擦焊

14 Welding Techniques for Manual Welding

Metal-arc welding electrodes are made with core-wire diameter from 2.0mm to 10mm. For all but exceptional circumstances, however, the useful range is 2.5 ~ 6.3mm. The length of each electrode depends on the diameter, for small-diameter electrodes where manipulation of the electrode calls for the greatest control the electrode may be only 300mm long. Generally, however, electrodes are manufactured to a length of 450mm and are consumed at a burn-off rate of 200 ~ 250mm/min. At one end of the electrode the coating is removed during manufacture so that it may be gripped in the electrode holder through which the welding current is introduced.

The working current range for a number of electrode sizes is indicated in Fig. 14-1. For economic reasons the welder should use the largest diameter of electrode suitable for each application. Because positional welding requires precise control of a small weld pool, smaller sizes of electrode are used for this purpose than for *downhand welding* where the pool is shaped by gravity. In multi-pass welds in fillets or grooved joints the first pass is usually laid using a smaller-diameter electrode to obtain better access and root penetration.

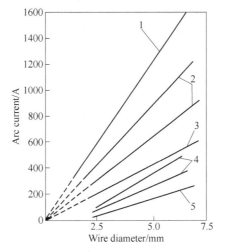

Fig. 14-1 Working ranges of electrodes for manual metal-arc and automatic process of welding mild steel

1—upper limit submerged arc; 2—continuous covered electrode plus CO_2; 3—continous covered electrode; 4—manual metal arc iron-powder type; 5—manual metal arc covered electrode

The welder's task is to direct the arc into the joint so that metal is deposited where required and to manipulate the electrode by whipping or weaving so that the arc force holds the metal in position and sweeps aside the *slag*. An electrode is never held perpendicular to the joint but is usually inclined so that an angle of about 110° is subtended between the weld bead and the electrode. This is sufficient to allow the welder to see the crater beneath the arc and for the arc force to prevent the undesirable flow of slag ahead of the advancing crater. As the process is intermittent a welder will usually have to stop and fit a new electrode many times in the course of making each electrode the weld crater pipe. To avoid this the arc is broken by drawing the electrode slowly back along the bead while at the same time lengthening the arc. Before the next electrode is used, the slag which has solidified over the crater must be chipped away to avoid slag inclusions. The arc from the new electrode is struck ahead of the crater and moved back to take up the end of the previous source of slag inclusions, porosity and lack of fusion. This part of the welding technique must be mastered thoroughly if the welder is to produce quality work.

Current, voltage (arc length) and speed are important process variables. Low current will give and irregular *weld bead* which sits on top of the plate; a high current will give irregular beads with poor penetration and a tendency to *slag inclusions*; high voltages and long arcs result in spatter and a tendency to pick up nitrogen from the air giving porosity. High welding speeds result in peaky, undercut beads; low welding speeds cause broad beads which tend to overlap on the work.

When welding steels of higher carbon equivalents it is often necessary to control the energy input per unit length of weld. The energy input per unit length, usually expressed in kJ/mm, together with the heat-sink effect (and preheat if used) determines the rate of cooling of the deposit and through this the microstructure and hydrogen diffusion in the joint. A low heat input, other things being equal, increases the risk of hydrogen cracking in the heat-affected zone while too high a heat input can affect the notch toughness of joints in certain steels adversely.

Energy input is the product of current and arc volts divided by welding speed. Because of the dependence of electrode melting rate on current and geometrical relationship: between core wire, melting rate and diameter on the one hand and deposit volume and welding speed on the other, however, the practical control of energy input is based on electrode diameter and run-out length. Run-out length is the length of deposit per electrode. Alternatively, for single-run fillet welds the leg length of the fillet can be specified.

14.1 Operating Characteristics of Manual Metal Arc (MMA) Welding

14.1.1 Welding Current

Either direct or alternating current can be used for MMA welding. To some extent, the choice is based on experience and individual preference, but a number of factors need to be considered:

(1) Virtually all MMA electrodes work on D. C., but only certain flux compositions give stable operation with A. C.

(2) Transformers are easier to maintain than the generators or rectifiers used for D. C. Also, A. C. units are more robust.

(3) D. C. arcs may be deflected from the joint by the magnetic effects produced when the welding current flows through the work (Fig. 14-2). This phenomenon is known as and is less co-

mmon with modern electrodes than it used to be. It can, however, sometimes lead to difficulties. Arc-blow does not occur with A. C. , as stable magnetic fields are not established.

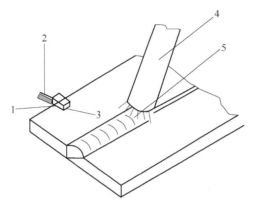

Fig. 14-2 Arc blow in MMA welding with direct current

1—connection to power supply unit; 2—welding return; 3—current flow; 4—electrode;
5—arc deflected along line of current flow to welding return

(4) Higher open-circuit voltages are required for A. C. The arc is extinguished each time the current goes through zero as the polarity is reversed (i. e. every one-hundredth of a second), Fig. 14-3. If the weld pool is to remain molten, the arc must be instantaneously reignited. This requires that a voltage in excess of 80V is applied to the electrode each time the current falls to zero. These high voltages can constitute a safety hazard and D. C. , with its lower o. c. v. of about 60V, is often preferred for this reason alone. It is expected that modified power-supply units and new types of flux covering will be available in the future to enable A. C. to be use without the need for high voltage.

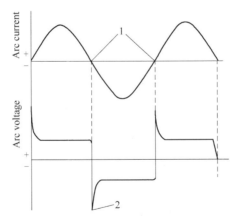

Fig. 14-3 Current and voltage wave forms in A. C. welding

1—arc extinguished as current goes through zero; 2—voltage tries to rise to o. c. v. value (this high voltage restrikes arc)

Ideally, the work should be positioned during welding so that the molten weld metal is held in place by gravity. This is called the *flat position* (Fig. 14-4) and give the welder the most favourable conditions for controlling the weld pool. It also enables high currents to be used, leading to faster welding. This implies that the work can be turned or manoeuvred easily. Many fabrications do not lend themselves to this treatment, and much of the welding in industry is done in position.

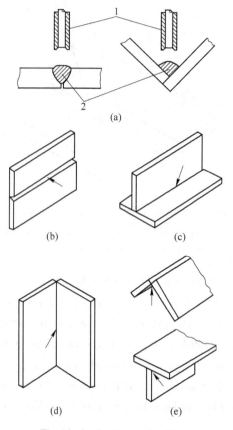

Fig. 14-4 Positions of welding
(a) flat; (b) horizontal; (c) horizontal vertical; (d) vertical; (e) overhead
1—electrode; 2—weld metal held in place by gravity

14.1.2 Position of Welding

Three main positions, in addition to flat, can be identified: horizontal, vertical, and *overhead*. There is subdivision of the horizontal position, known as the horizontal-vertical, which relates specifically to T joints in which the axis of one member is vertical while the other is horizontal. In all these positions the metal tries to run out of the joint under the effect of gravity, and the welder's technique must be modified to combat this tendency. A major contribution is made by the flux, and this will be discussed in more detail later. The welder controls the weld by lowering the heat input to reduce the fluidity and to give a small pool which solidifies before it has time to run out of the joint. At the same time, the direction of the arc, i.e. the angle between the electrode and the weld surface, can be varied to position the weld pool to the best advantage.

The maximum current is lower in positional welding. Whereas 350A can be readily used for joints in the flat position, the welder would have considerable difficulty in working above 160A when overhead. It follows that the sequence used to deposit a weld of a given size differs from one position to another. This can be illustrated by considering the deposition of a typical fillet weld. The size of a fillet weld can be specified in a number of ways. But for a present purpose we will specify a leg length of 10mm. This is the distance from the root of the weld to the toe and is, in effect, one side of the triangle formed by the fillet (Fig. 14-5).

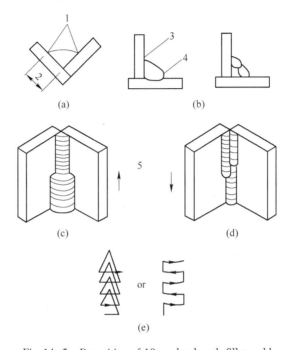

Fig. 14-5 Deposition of 10mm leg length fillet weld
(a) flat one pass; (b) horizontal vertical; (c) vertical up (two pass);
(d) vertical down (three pass); (e) electrode movement for second run (weaving)
1—toes; 2—leg length 10mm; 3—undercut; 4—overlap; 5—direction of welding

In the flat position, with a current of 300A the weld could be deposited in one pass. The welder moves the arc from side to side to ensure that the faces of the joint are properly fused. This is known as weaving and it produces a smooth flat surface. Using this technique, a one metre length of weld could be deposited in about 10min.

If the joint is in the horizontal-vertical position, attempts to deposit the weld in one pass result in a large uncontrollable pool which tries to flow on to the surface of the horizontal plate. The resultant weld is *misshapen* and there is deposited in three runs, using a lower current of about 200A. Although the travel speed for each individual run is faster, the total time for the joint is 15min, i. e. 50% longer than in the flat position.

When the joint is positioned so that the weld is deposited overhead, the current is reduced to 160A and four or five runs may be needed to obtain the required size. The time has now increased to about 24min.

When the plates are arranged so that the joint line is vertical, the weld can be started at the bottom and the arc moved upwards. This is the conventional method of fillet welding a joint in this position and is often referred to as the vertical-up technique. Two runs are required at about 145A and the welder weaves the electrode along a triangular path to achieve fusion of the joint faces. The weaving also helps to distribute the heat and control the weld-pool fluidity. If the welder experiences difficulty in preventing the surface from *sagging* at the centre-line, the current is reduced to about 120A and the weld is deposited in three passes using a smaller weave. The weld times range from 20min to 25min, depending on the current used.

Alternatively, a *vertical-down* technique can be used. Welding starts at the top of the joint and the electrode is pointed upwards at the deposited weld metal. The rate of travel is critical, as the molten metal must not run down the joint ahead of the arc-if it does, fusion of the parent metal may not take place. The weld pool must be relatively small, and there is little scope for weaving. This means that our 10mm leg length fillet weld may need five or even six weld runs. On the other hand, travel speeds with the vertical down technique are comparatively high, and the overall time for the joint is about 17min/m. The two main drawbacks with this method of welding vertical joints are that there are only a small number of suitable electrodes and it requires considerable skill to produce welds which are free of lack of side fusion.

14.1.3 Functions of Flux Covering

The main reason for using a flux covering in MMA is to protect the molten metal from atmospheric contamination. At the same time, the flux fulfills a number of functions, all of which contribute to the success of the welding operation.

14.2 Weld Metal Protection

The flux melts in the arc, along with the core wire. It covers the surface of the molten metal, excluding oxygen and nitrogen. When it has solidified, the flux forms a slag which continues to protect the weld bead until it has cooled to room temperature. While it is important to ensure that a good flux/slag covering is achieved, the flux composition must be chosen with a view to slag *detachability*, as the slag must be completely removed before the next run is laid. Ideally we would like a slag which lifts off the completed weld by itself. This is difficult to reconcile with the need for the slag to adhere to the weld metal during the cooling period, to avoid entry of air which would oxidize the covering to achieve other objectives. The result is always a compromise, and the ease with which the slag can be removed varies from one electrode type to another.

Additional protection against atmospheric contamination can be provided by including compounds which decompose in the heat of the arc to form gases. These replace air in the arc atmosphere, thus reducing the risk of oxygen and nitrogen absorption. They may be carbonates giving off carbon dioxide, or cellulose which produces an atmosphere of hydrogen and carbon monoxide.

14.3 Arc Stabilization

Although we often talk of the need to ensure that the arc is stable, it is very difficult to define what we mean by this. From an operational point of view, the easiest explanation is that we want the conditions to remain constant along the joint, unless the welder decides to make a change to accommodate some variation in fit up, etc. This implies that the top of the arc should always be at the centre of the core wire cross section, and the arc column should be in line with the axis of the electrode. In a groove, such as a single V joint, the arc must not move from the electrode to the work; it must stay firmly fixed in the direction dictated by the welder. At the same time, the end of the electrode must melt uniformly and the metal must be transferred to the weld pool without upsetting the stability of the arc.

Another aspect of stability is the ease with which the arc can be established at the start of a weld run or reignited at the beginning of each half cycle when using A. C. In both cases, the gas in the arc gap must be ionized rapidly and at the lowest possible voltage. Ionization is facilitated by inclusion of titanium oxide, potassium silicate, and calcium carbonate in the flux. Silicates and oxides which have been added principally for other reasons can also act as arc stabilizers.

14.4 Control of Surface Profile

To understand the action of the flux in controlling the profile of a weld bead, we must first look at the role played by the surface tension of the weld pool. If the surface tension is high, the molten-metal surface becomes *convex*. On a flat plate, the liquid pulls away at the edges, and the angle of contact between it and the solid surface approaches 90°. This is another way of saying that the molten weld metal does not wet the solid parent metal. At the other end of the scale, if the surface tension is low, the contact angle is small and the molten metal has good wetting characteristics, so the surface of the pool is very flat.

Neither of these extremes is desirable in welding. A very high surface tension not only gives poor profile but also prevents the metal from flowing uniformly in the root of a V groove. On the other hand, a very low surface tension makes it difficult to control the profile and to restrict the size of the pool. There is also a risk that the weld metal will follow over the joint faces before the arc has had time to melt them. Once again a compromise must be reached for satisfactory practical use.

The surface tension of the pool in arc welding is controlled by the oxygen level of the weld metal. This, in turn, is determined by the oxygen content in the flux. If the oxygen in the weld is low, this element will be transferred from the flux until a stable value has been established. The more oxygen there is in the flux, the higher will be the level in the weld. The effect this has on surface tension, and hence on surface profile, can be seen by examining fillet welds deposited in the horizontal-vertical position (Table 14-1). A low-oxygen weld has a high surface tension, and vice versa.

Table 14-1 Effect of flux oxygen content

Oxygen content	High	Medium	Low
Surface tension	Low	Medium	High
H-V fillet weld profile			

In some cases, surface profile is not a prime consideration in formulating a flux, and a less-than-satisfactory profile may have to be accepted. This is the case with fluxes containing appreciable amounts of calcium *fluoride* and carbonate. These are used to give better mechanical properties which are associated with low oxygen contents. As we have seen, however, this means that the surface tension is high and we will inevitably have a convex profile if we use electrodes covered with these fluxes.

14.5 Control of Weld Metal in Position

The slag can be used as a mould which helps to keep the weld metal in place while welding in position. Three physical properties of the liquid slag must be kept in balance. Firstly, it must have sufficient fluidity to flow freely from the root of the weld to give good visibility and to avoid slag being trapped when the weld solidifies. The fluidity must not be too high, however, otherwise the flux runs off the face of the weld. The problem is eased if the surface tension is reasonably high, since this helps the slag to stay in place. Finally, the slag should solidity rapidly to forms solid barrier which remains the tendency for the weld metal to run out of the joint.

14.6 Control of Weld Metal Composition

One of the outstanding advantage of MMA welding is that we can make adjustments to the composition of the weld metal by adding alloying elements to the flux covering. We have already noted that the oxygen content of the weld pool depends to a large extent on the amount of this element present in the flux. Similarly, if we add manganese to the flux, in the form of ferro-manganese, it will transfer into the weld. The actual amount which ends up in the weld depends partly on the concentration of manganese but also on the composition of the flux. For any given composition of flux and weld metal, alloying elements are distributed between the two in a more or less fixed proportion. The transfer can, of course, work both ways. If the flux or slag layer is low in manganese, this elements transfers from the weld until the correct proportion is established. Thus elements can be both added to and taken from the weld simply by altering the flux composition.

A good degree of control can be exercised over this transfer mechanism. The thickness of the covering on the electrode can be maintained to within close tolerances. Hence the ratio of flux to core wire is reproducible and the amounts of the covering on the electrode can be maintained to within the amounts of alloying elements which need to be added to produce a particular weld metal composition can be calculated by the electrode manufacture.

Broadly speaking, there are three main aspects of weld metal composition control which should be considered: alloying, deoxidation, and contamination.

Words and Expressions

downhand welding   平焊
slag   熔渣
weld bead   焊接熔滴
slag inclusions   焊接中的夹渣
flat position   平焊位置
overhead   仰焊

misshapen   畸形
sagging   下垂
vertical-down welding   向下立焊
detachability   脱渣性
convex   凸出的
fluoride   氟化物

15　Gas Shielded Arc Welding

15.1　Historical Background

The idea of using a *gaseous shielding medium* to protect both the *electric arc* and weld metal from contamination by the atmosphere is almost as the covered electrode. Roberts and van Nuys in 1919, and others several years later, considered the problem and a variety of gases were proposed from the inert gases to hydrogen and *hydrocarbons*. In the 1930s the interest began to centre on the *inert gases* but it was not until 1940 that experiments were begun at the Northrop Aircraft Co. of USA with the deliberate intention of developing a practical inert-gas welding method. The metal to be welded was melted by an arc struck from a tungsten electrode, in an atmosphere of the inert *monatomic gas* helium.

The original apparatus comprised the simple tungsten electrode touch and a D.C. generator. Arc starting was by brushing the electrode on the work but this led to contamination of the electrode and a high frequency spark generator was added to the equipment so that an arc could be struck from the electrode without touching it on the work. At first both electrode negative and electrode positive polarity were used, although the negative polarity was favoured because less heat was generated at the tungsten electrode, which remained relatively cool.

With the desire to weld thicker material, welding currents of over 100A became necessary and it was no longer possible to use the electrode positive polarity because the tungsten electrode became so hot that molten tungsten dropped off into the weld pool. The higher welding currents also necessitated water cooling of the body of the torch because of the increased amount of heat conducted back along the electrode.

By 1944 it was recognized that electrode polarity was of greater significance than had appeared at first. Up to that time the inert gas arc process had been used principally on thin gauge magnesium and stainless steel, but attempts had also been made to weld aluminium with which it was found necessary to employ a flux. It was observed, however, that oxide removal could be accomplished by the arc itself on electrode positive D.C. or in A.C. welding, thus making a flux unnecessary. Unless a certain minimum open circuit voltage was available when welding aluminium with A.C., the oxide film was not broken down so that the A.C. was rectified and welding was impossible. By 1946, however, it had been found that the spark ionizer could be made to stabilize the A.C. arc. Gradually a preference emerged for argon over helium in manual welding, largely as a result of the smaller change in arc voltage with arc length when welding with argon. This made the process less critical from the welder's point of view.

Once a start has been made to the welding of aluminium by the inert-gas tungsten arc method there began a period of rapid development because of the new range of applications opened up. Although limited for several years to welding sheet material at less than 150A there was now a demand to go to higher currents. Metal gas *nozzles* were replaced by ceramic ones, these in turn being replaced by water cooled metal nozzles when it was found that the ceramic nozzles had a limited life. The water cooled torch body and power lead was now essential to give lightness and flexibility to the torch and, because the high frequency ionizer was left on continuously, great attention had to be paid to insulation.

Although the high-frequency ionizer stabilized the arc it did not affect the inherent unbalance between the voltage on alternate half-cycles which resulted in a D. C. component that tended to *saturate* the transformer. At first this was overcome by applying a similar D. C. voltage, but of opposite polarity, to the circuit so that the D. C. component was balanced out. This was done with storage batteries, but subsequently it was found that large capacitors in series with the arc had the same effect.

The purity of the shielding gas was improved from 98% to over 99.95% as the process developed, particularly as a result of the need for high purity gases for the welding of aluminium alloys and reactive metals. Argon, the only inert gas available outside the USA, gained in popularity even in that country, being the chief gas used for manual welding, although the higher arc voltage and, therefore, greater penetration of the helium shielded arc was found of value in automatic welding. Both helium tungsten arc and argon tungsten arc techniques were rapidly applied to the welding of a range of non ferrous metals which had proved difficult to weld by other methods.

With aluminium, as with magnesium, the new process gave greater scope to the engineer because of the absence of flux. Previously fillet welds and other types of joints in which flux might be trapped had to be avoided because of the danger of corrosion after welding. The more concentrated heat input of the tungsten arc welding process over gas welding enabled welding speeds to be increased and improved the metallurgical quality of welds. Although there are many advantages to the process it was also found to have limitations. The separate addition of filler metal required the use of both the welder's hands and, therefore, access to difficult joints tended to be restricted and positional welding was slow and difficult.

In 1948, however, the second important gas shielded process made its appearance and was to prove capable of being used satisfactorily on many of the types of joint which were not ideally suited to the tungsten arc method. In tungsten arc welding the electrode was non consumable, but in the new method the electrode was in the form of wire which was consumed during welding to provide filler metal for the weld. This wire was fed from a coil to the arc at the same rate as it was melted away. The term metal arc is used to describe an arc welding process in which the electrode is consumed during welding to provide filler metal for the weld and the new process therefore became known as inert gas metal arc welding. It was not long before gases other than inert were used so that the process should now strictly be described as argon metal arc, helium metal arc or CO_2

metal arc, etc., as appropriate, with the general title of gas metal arc for the whole series.

In the first apparatus the wire was pushed through a flexible tube to a *pistol* type torch where contact was made with the welding current conductor. Argon gas to shield the weld pool was passed through a nozzle surrounding the filler wire. Although the torch was held in the hand the process possessed certain characteristics usually associated with automatic welding. It was the first manual process to utilize the principle of the self adjusting arc in which the arc length is held constant during welding, irrespective of movement by the operator. An essential feature of the process, which made it possible to use both a self adjusting arc and the flexible feed tube to the torch, was the small diameter of the electrode wire, usually about 1.6mm. Metal was transferred axially from this wire electrode to the work in a stream of fine drops.

Development of the inert gas metal arc method in the early 1950s was closely associated with the welding of aluminium alloys which at that time were becoming established as structural materials, in particular for shipbuilding where a process was needed which would weld in any position. Had the need for structures in aluminium alloys not existed in 1950 the process might well have been developed more slowly, and it was fortunate that aluminium was one of the first metals to be tried for, as is now known, metal is transferred across the arc more satisfactorily with aluminium than with any other metal.

Following the successful use of inert gas metal arc welding with aluminium, attempts were made to apply the method to other non ferrous metals and to steels. The use of argon for welding steels was not economically attractive at that time but, after several years of research in the USSR, UK and USA, techniques were developed which permitted the satisfactory use of carbon dioxide as a shielding gas. This gas is cheap and made the process competitive in many applications with established processes such as metal arc. More recently argon based mixtures have become popular.

The history of gas shielded arc welding, from the first use of the helium tungsten arc in the 1940s to the successful use of the carbon dioxide shielded metal arc in the 1960s, has been discussed in reasonable detail because this is possibly the best introduction to this important series of processes. The impetus behind each new development can be seen in perspective and it will have been noted that the circumstances have been extraordinarily favorable for rapid exploitation.

15.2 Inert Gas Tungsten Arc Welding

Although electrodes of *refractory metals* other than tungsten have been used for inert gas welding they are unsuitable because they erode too easily. Even tungsten electrodes are eroded, but the rate with careful usages so low that the electrode is justifiably considered non consumable. The gas which surrounds the arc and weld pool must also protect the electrode. At the high temperatures reached at the root of the arc the tungsten is readily oxidized so that the shielding gas can only comprise mixtures of the inert gas can only comprise mixtures of the inert gases and hydrogen, or in extreme cases nitrogen. Hydrogen is not generally useful for shielding because it raises the arc voltage and requires a high open circuit voltage and it can also be absorbed by some metals giving rise to either cracking or porosity. For practical purposes, therefore, gas shielded non consumable electrode welding employs argon or helium for shielding and tungsten for the electrode.

15.3 Electrode Polarity

Because of the greater heat liberated at the anode a tungsten electrode used on this polarity becomes more readily overheated than if it is the negative pole of the arc. The maximum current which the electrode will carry is reached when the molten end becomes so large as to be unstable and particles of tungsten begin to leave the electrode. Even with tungsten electrodes of 6mm diameter, no more than 100A can be used with D. C. when the electrode is positive. When the electrode is negative, however, the permissible current is up to eight times greater. For this reason tungsten arc welding with the electrode positive is seldom used.

The chief advantage of the D. C. electrode positive method is the cleaning action exerted by the arc on the work. It is by no means certain that the widely held theory that is because of *ion bombardment* is entirely correct. High speed films of the arc indicate wide and exceedingly rapid motion of the *cathode spots* which have a preference for particles of oxide and other impurities. Vaporization of both oxide and underlying metal could occur at these spots, any oxide remaining is broken up and freed to float away to the edges of the weld pool. This activity of the cathode spots can sometimes be observed, particularly with aluminium, on the edges of the plate adjacent to the weld pool. Except where the cleaning action is essential, when welding aluminium or alloys containing appreciable amounts of elements forming refractory oxides, the D. C. electrode positive polarity is normally used.

The characteristic motion of cathode spots is often said to cause instability when the electrode is negative because the root of the arc can wander over the end of the electrode. Two measures are adopted to prevent this: the electrode is ground to a taper and the tungsten is doped with materials to improve its *emissivity*. The doping of electrodes with 1% ~ 2% *thoria*, or with *zirconia*, increases the area of the cathode spot and also gives easier arc starting and improved resistance to contamination. Contamination and loss of tungsten occur largely at the start of a weld but by improving the emissivity hot spots are avoided and the electrode is made to achieve its operating temperature more easily. An additional advantage is that the current carrying capacity of the electrode is raised.

The angel of the cone which is ground on the end of the tungsten electrode affects the penetration shape of the weld metal. It is believed that a plasma jet can develop from electrodes with a small cone angle giving deep narrow penetration when welding conditions are such that the full thickness of the workpiece is penetrated. On thick plate when there is partial penetration the plasma jet is deflected causing a wide bead at the surface. With a large cone angle the plasma jet does not develop and thick plate can be welded with a narrow bead.

Where an automatic arc length control system is employed it is important to use a consistent cone angle because the arc voltage arc length relationship is influenced by cone angle. A tolerance of

±5° in cone angle should give an arc voltage consistent to ±1V and this will be adequate for most purposes. Spiller and MacGregor recommend that cone angles should be chosen according to applications as follows:

(1) Full penetration welds using 50~200A (maximum penetration for lowest current) —cone angle 30°~60°;

(2) Full penetration welds using 50~200A (where tolerant welding conditions are necessary) —cone angle 90°~120°;

(3) Filler runs using less than 200A—cone angle 60°;

(4) Filler runs using more than 200A or fillet (welding with minimum undercut) —cone angel 120°.

See also the effect of cone angle when using pulsed tungsten arc welding.

Where the material being welded demands the electrical cleaning action of the electrode positive polarity, but currents over 100A must be used, an A. C. power supply is employed. The A. C. arc combines the advantages of arc cleaning of the work on the half cycle in which the electrode is positive with the lower heat input to and, therefore, cooler running of the electrode when it is negative. When used on A. C. the end of the electrode is not tapered and should assume a stable hemispherical shape because of superficial melting. If the current is excessive for the size of electrode this molten tip will *oscillate* because of the *pulsating arc* forces and particles of tungsten may be ejected from a small pip which forms in the centre. Too low welding current may not provide sufficient energy to melt the electrode end so that the arc root wanders making the arc unstable. The optimum current range for each size of electrode depends on of the electrode, the type of electrode and, possibly most important, the balance between the positive and negative parts of the current cycle (the convention for reference to polarity is with respect to the electrode). As will be seen later, there is a tendency for the negative current half cycle to be large than the positive, giving a D. C. component. If this is eliminated to protect the transformer from saturation the positive half cycle is increased in duration so that the electrode runs hotter and its current carrying capacity is reduced.

15.4 *Arc Maintenance*

With A. C. the reversals of voltage and current introduce the problem of arc reignition as the arc is extinguished twice in every cycle. When the electrode becomes negative the arc ignites satisfactorily but when the voltage is reversed so that the electrode becomes positive the arc goes out and will not reignite unless at that instant there is sufficient voltage available at the arc gap. This problem is similar to that met in A. C. metal arc with the difference that the voltage required is much higher for the negative positive change with the tungsten/aluminium arc. There is also a greater contrast between the case of the electrode positive/negative change and the subsequent reversal. This is indicated by the marked *reignition peaks* which are observed on the voltage records of the tungsten aluminium arc.

The reignition can be accomplished satisfactorily in three ways:

(1) With a well designed transformer with low electrical inertia the voltage required for reignition can often be supplied by the transformer giving the process of self reignition. If a high frequency spark unit is used for striking the arc this can then be switched off by a relay once welding begins. Alternatively, the high frequency spark unit may be operated from the open circuit voltage so that it ceases work when the voltage drops to that of the arc. One of the advantages of switching off the high frequency spark is that the radio interference caused by the sparks is limited in duration.

Self reignition for all its simplicity has disadvantages. The open circuit voltage required tends to be high, usually approaching 100V, and the power factor has to be low because a high voltage must be available at current zero. Greater reliability is claimed for the methods in which means are specially provided for assisting reignition.

The voltage and current waveforms show the cycle of events in arc reignition. If the voltage across the gap climbs rapidly towards the open circuit voltage the reignition voltage for the arc is quickly attained and the arc is restarted.

A delay in restriking leads to a discontinuity in the current waveform called a current zero pause. In extreme cases, where, for example, there is a reduction in open circuit voltage and therefore the available *restriking voltage*, the arc may not reignite at all on the positive halfcycle. This is complete rectification, a condition quite unsuitable for welding as there is no cleaning action and the transformer can be overloaded by saturation with the D. C. The delayed reignition is known as partial rectification.

(2) When left on continuously, the high frequency spark unit used for spark starting can become an arc maintainer. Reignition is accomplished by the sparks which discharge across the arc gap providing an ionized path for the main power circuit. Slightly lower open circuit voltages are required with high frequency reignition. The spark unit comprises a capacitor, charged by a high voltage transformer, which discharges through a spark gap. A train of sparks is set up which lasts as long as the transformer voltage exceeds the breakdown voltage of the spark gap. This generally occurs for two thirds of each half cycle of the mains supply. The unit, therefore, emits bursts of sparks which must be arranged to occur during the period in which the welding circuit passes through current zero. Because of the cyclic nature of the sparks reignition can not be instantaneous and there is some partial rectification.

(3) In the third method of arc maintenance a voltage surge is injected into the power circuit to supply the reignition peak. This is done by discharging a capacitor through a switch which is tripped automatically by the power circuit. When the arc is extinguished at the end of the negative half cycle the reignition peak begins to develop and itself fires a gas discharge valve which discharges the capacitor. Reignition is therefore instantaneous eliminating partial rectification, and because the transformer does not have to supply the full reignition voltage it is possible to reduce the transformer open circuit voltage. Welding can actually be accomplished at less than 50V which permits an improvement in power factor as well as giving greater safety. The timed surge is an arc maintainer only and will not strike an arc from the cold or always after momentary extinction. A

spark generator operated by the surge is used, therefore, to start the arc and is switched out automatically once the arc is started.

15.5 Direct Current Component

Mention should now be made of the inequality of the half cycle loops in the current waveform. The positive half cycle tends to be the smaller because the arc voltage when the electrode is positive is higher than when it is negative. This is because the sum of the cathode and anode drops is different when the cathode is on the work from when it is on the electrode. The effect, which is known as *inherent rectification*, is independent of reignition phenomena and is observed even when the arc is running correctly. It results in a D.C. component of current which tends to saturate transformers making it necessary to *derate* them to about 70% of their usual value. At high welding currents it can also cause serious arc blow. Any current zero pause also causes current unbalance and adds to that caused by inherent rectification.

These effects are usually considered *detrimental* so that inherent rectification is corrected by the insertion of a large capacitor in the power circuit. Banks of reversible *electrolytic* capacitors giving about $100\mu F/A$ are used. A charge is left on these capacitors when the electrode is negative which is available when it becomes positive to increase the voltage available and therefore to drive more current through the arc. The D.C. component can also be suppressed by placing large storage batteries in series with the arc so that the D.C. component is biased out.

Inherent rectification or the D.C. component has an important thermal effect in welding. Because it is the result of lack of balance between positive and negative half cycles it affects the distribution of heat between work and electrode. The electrode runs hotter when the inherent rectification is balanced out and the current on the positive half cycles is increased. Conversely, more heat is available in the work when there is no D.C. suppression. With weld beads made at the same current setting and speed, it will be noticed that those where D.C. suppression is used can be 15% narrower. In the USA where D.C. suppression is less common than in Europe the recommended welding conditions tend to higher currents for a given size of electrode and faster welding for the same current.

15.6 Starting the Welding Arc

Contamination and tungsten loss can be greatly reduced by striking the arc with a low current through a *pilot circuit* and switching to the main current a few cycles later when the electrode has warmed up. In the absence of such techniques, and whether touch starting is used or not, it is helpful to strike the arc on a piece of scrap material and then restrike on the work once the electrode has reached operating temperature. These measures are particularly important when welding at high currents because the chances of losing tungsten to form tungsten inclusions are greatly increased during the warming up period of the electrode.

During the warming up of the electrode on A. C. and until it reaches a temperature which will permit *thermionic emission* it is possible for the arc to fail on the electrode negative half cycles. This leads to rectification in the opposite sense to that normally encountered while the arc is running so that the occurrence is known as inverse rectification. Once the electrode is hot the arc reignites readily with its cathode on the tungsten and it is more likely to extinguish when the work is negative. The most important effect of inverse rectification is that it leaves a charge on the D. C. suppression capacitor of the reverse polarity to normal. This opposes the half cycles of electrode negative current which are required to heat the electrode, so that the arc puts itself out to prevent this, the suppressing capacitors are often switched out when the arc is being struck and switched in later when it is running normally.

15.7 Welding Techniques

After the arc has been struck the torch is held stationary while the molten weld pool is formed. If the welding current is adequate this should not take more than a few seconds and the surface of the pool should be bright and clean. A leftward technique is used with the torch held at 80° to give visibility of the pool and to sweep argon ahead of the bead. Once the pool is established welding can proceed and filler metal may be added if required. The action of introducing filler metal to the weld pool can disturb the gas shield and entrain air. It is helpful, therefore, to keep the end of the filler rod within the shield at all times. This also prevents oxide forming on the end of the filler rod.

If the filler rod diameter is too small in manual welding it will melt rapidly, forming a globule at its end. Conversely, too large a rod may disturb the arc and cause oxide inclusions by shielding the weld from the arc cleaning action. Filler rod of the correct size can be held close to the pool and no violent motions are necessary when adding metal. With mechanized welding, filler wire from a coil is fed into the leading edge of the weld pool and is arranged to make good contact with the solid metal just ahead of this edge. Smaller diameter wires can be used than for manual welding where this contact can not be maintained.

Mechanized welding is widely used with the tungsten arc and frequently to simplify the process the joints are designed so that the addition of filler metal is unnecessary. This requires the use of edge joints or accurately machined and fitted butt joints. When the latter are used it is usual to allow the weld bead to sink slightly into a backing bar or under gravitation so that the weld occupies the full cross section of the joint, although no extra metal is gathered into the pool, increasing the thickness in the weld bead and leaving transverse shrinkage or deformation in the welded component. Although the addition of filler metal is avoided if possible it is sometimes necessary so that the composition of the weld metal can be changed to avoid cracking or porosity.

Equipment for mechanized tungsten arc welding frequently incorporates an arc length control device. The torch is mounted on a linear actuator which responds to arc voltage and this enables the torch to follow an uneven surface while still maintaining a constant arc voltage and hence a constant arc length. Arc length control was originally employed to permit the torch to be traversed along a weld seam of varying contour or to allow an unsymmetrical component to be revolved under a stationary welding head. For this purpose actuators with strokes of up to 300mm or more are employed. A more frequent use of arc length control. However, is to accommodate minor unwanted variations in the workpiece which arise from eccentricity in pipes or the bowing of longitudinal jigs as a result of clamping pressure or thermal strains. Another excellent application is in the tungsten arc welding of pipe rotated under a torch when a root run and several filling passes, the latter requiring filler metal, are required. The arc length control automatically lifts the torch as the joint is filled to maintain a constant arc length. If the arc length (and therefore arc voltage) was allowed to vary there would be an accompanying change in arc current and consistent fusion would be difficult to maintain.

15.8 Stopping the Weld

When the current is cut off and the arc extinguished the electrode and weld pool begin to cool, but an adequate flow of argon must be maintained until the danger of oxidation is past. In most tungsten arc welding equipment the flow of gas is controlled by a *solenoid valve* which is timed to open before the arc ignites and to close after allowing sufficient time for the electrode to cool.

When the arc is extinguished abruptly at full welding current the weld pool solidifies with a central pipe, a defect which can cause leaks in joints intended for service in vacuum or under pressure. The crater pipe is overcome by reducing the current gradually before switching off with a device known as a crater filler. In this way the pipe is fed with liquid metal and the crater solidifies progressively. It is possible to reduce the crater in mechanized welding by increasing the speed before switching off. With manual welding the welder should feed the crater with filler metal. In D.C. welding the arc may be extinguished by lengthening the arc gap, causing the voltage to rise and the current to drop prior to extinction in a manner depending on the volt amp characteristics of the power source.

15.9 Applications

The tungsten arc process is used with welding currents from 1/2A up to 700A or 800A and is one of the most versatile methods of welding. The lowest currents are used with delicate air cooled torches to weld metal 0.05mm thick. For these very low currents DCEN is usually used A.C. is employed between 25A and 350A, but above this current the risk of tungsten inclusions increases. Up to 800A, DCEN is again used and these high currents are employed particularly for welding thick copper. Torches designed for up to about 100A are entirely air cooled but for higher currents water cooled torches must be used.

Although high welding currents permitting the welding of thick metal are possible, tungsten arc welding is primarily a process for welding sheet metal or small parts. The process is at its best when welding single pass or double sided close butt joints, edge joints or outside corner joints. It is less suitable for fillet welds with which care must be taken to obtain good fusion into the root. Because it is so easily mechanized and gives high quality welds, the process is greatly favored for precision welding in the aircraft, atomic energy and instrument industries. Circumferential and edge welds, for example, can sealing joints, are very suitable for mechanized tungsten arc welding. Arc length control systems are sometimes used with mechanized welding in which the arc voltage provides a signal to raise or lower the welding heat so that the arc can follow a curved or undulating surface.

Various automatic devices are available for welding tubes to tubeplates. Typical of these is the method in which a miniature torch revolves around a central *spigot* which is fitted into the tube.

Words and Expressions

gaseous shielding medium　气体保护介质
electric arc　电弧
hydrocarbons　碳氢化合物
inert gases　惰性气体
monatomic gas　单原子气体
nozzle　喷嘴
saturate　浸透，饱和
tungsten arc welding　钨极电弧焊
pistol　枪式焊接器
refractory metal　难熔金属
ion　离子
bombardment　轰击
cathode spot　阴极斑点
emissivity　辐射率
zirconia　锆
thoria　钍
oscillate　振动，摆动
pulsating arc　脉动电弧
arc maintenance　维弧
reignition peak　再起弧峰值
restriking voltage　再起弧电压
derate　降低
detrimental　有害的
electrolytic　电解的
pilot circuit　控制电路
thermionic emission　热电子发射
solenoid valve　电磁阀
spigot　套管

A new approach to quantitative analysis of bainitic transformation in a superbainite steel

Guang Xu,[a,b,*] Feng Liu,[a] Li Wang[b] and Haijiang Hu[a]

[a]*Key Laboratory for Ferrous Metallurgy and Resources Utilization of Ministry of Education, Wuhan University of Science and Technology, Wuhan 430081, People's Republic of China*
[b]*State Key Laboratory of Development and Application Technology of Automotive Steels (Baosteel Group), Shanghai 201900, People's Republic of China*

Received 1 January 2013; revised 28 January 2013; accepted 29 January 2013
Available online 5 February 2013

In situ observation of bainitic nucleation and growth in a superbainite steel was conducted by high-temperature laser scanning confocal microscopy. The morphological development of bainite transformation was directly observed. Moreover, thermal simulation tests were carried out in which bainite transformation was recorded by dilatometry. The bainite transformation process of a superbainite steel was quantitatively analyzed by a new approach that combines laser scanning confocal microscopy and dilatometry in a thermal-mechanical simulator.
© 2013 Acta Materialia Inc. Published by Elsevier Ltd. All rights reserved.

Keywords: Bainitic steels; Phase transformations; Microstructure; Isothermal heat treatments; Dilatometer

Bainitic transformation has been a topic of interest since Bain reported the first bainite in the 1930s [1–3]. There is also considerable interest in experimental methods to achieve bainitic transformation during isothermal holding. In recent years researchers have proposed a new method for bainitic transformation analysis, i.e. in situ observation by high-temperature laser scanning confocal microscopy (LSCM) [4–7]. Zhang et al. [4] studied bainitic transformation of a 0.15% C steel using LSCM. Kolmskog et al. [5] directly observed bainite formation below the martensite start temperature (Ms) by LSCM. Moreover, Yada et al. [6] studied the lengthening of bainitic plates in Fi–Ni–C alloys using a hot-stage microscope. Other in situ observations of bainitic transition were reported by Kang et al. [7], who investigated the nucleation and growth of bainite using in situ transmission electron microscopy observation. Additionally, Pak et al. [8] examined displacive phase transformation and surface effects on martensitic transformation with LSCM. Maalekian et al. [9] studied the austenite grain growth of a Ti–Nb microalloyed steel using a laser ultrasonics methodology in a thermal simulator. Zhang et al. [10] investigated the nucleation and growth of acicular ferrite in a Ti-added C–Mn steel by LSCM. The advantage of in situ observation is that the dynamic phase transformation process at any temperature and time can be continuously observed in real time, which provides a direct method to observe the bainitic nucleation and growth. The disadvantage of in situ investigation, however, is that bainitic transformation can only be qualitatively analyzed.

In this paper, a unique experimental technology, combining LSCM and dilatometry in a thermal-mechanical simulator, is successfully proposed to monitor the phase-transformation process in real time. The phase-transformation process and the morphological development of a Fe–C–Mn–Si superbainite steel are analyzed as an example of the combined observation. LSCM is utilized for in situ observation of bainite morphology during the phase transition, while dilatometry is used to quantitatively analyze the bainitic transformation process.

A superbainite steel with the chemical composition of 0.40 C, 2.81 Mn, 2.02 Si, balance Fe (wt.%) was used in the present study. The material was refined in a vacuum induction furnace and cast into a small ingot followed

---

* Corresponding author at: Key Laboratory for Ferrous Metallurgy and Resources Utilization of Ministry of Education, Wuhan University of Science and Technology, Wuhan 430081, People's Republic of China. Tel.: +86 027 68862813; fax: +86 027 68862807; e-mail: xuguang@wust.edu.cn

by rolling to a 10 mm thick flat. Samples for LSCM were machined to a cylinder of 6 mm diameter and 4 mm height. The top and bottom surfaces of samples were polished conventionally to keep the measurement face level and minimize the effect of surface roughness. The investigations were conducted on a VL2000DX-SVF17SP laser scanning confocal microscope. The specimen chamber was initially evacuated to 6×10^{-3} Pa before heating and argon was used to protect specimens from surface oxidation. The bainite start (Bs) and finish temperatures (Bf) and the Ms are 423, 305 and 256 °C, respectively, calculated by MUCG83 software developed by Bhadeshia at Cambridge University. Therefore, the isothermal temperature for bainitic transformation is chosen as 330 °C.

The specimens were heated at a rate of $5\,°C\,s^{-1}$ to two austenization temperatures (i.e. 1000 and 1100 °C) and held for 30 min. Then the specimens were cooled to 330 °C at a rate of $5\,°C\,s^{-1}$ and isothermally treated for 60 min for bainitic transformation followed by final air cooling to room temperature. The LSCM images were recorded continuously at 15 frames s^{-1} at 100× magnification during isothermal treatment at 330 °C. A video showing bainite transition process was simultaneously obtained. The ultrafine bainite can be formed in the material with designed composition after treatment using such technology. The principle for in situ observation of bainite nucleation and growth is the relief phenomenon occurring in phase transition. Ko and Cottrell [11] first observed the relief phenomenon of bainitic transformation in 1952.

In order to quantitatively analyze bainitic transformation, phase transformation experiments were carried out using thermal simulator with exactly same routes as used for LSCM. The cylindrical samples 8 mm × 20 mm were machined and tested on a Gleeble 1500 thermal simulator and the dilatation of specimens was recorded by dilatometry in a thermal-mechanical simulator.

Figure 1 shows the in situ observations of morphological development of bainitic ferrite during an isothermal holding process at 330 °C after austenization at 1100 °C for 30 min. The evolution of bainite sheaves is discussed in this work. It can be seen that nucleation of bainite takes place in the grain as shown by arrow A and at the grain boundary as shown by arrow B in Figure 1a. The secondary bainite ferrites nucleate sympathetically on the surface of the primary bainitic ferrite,

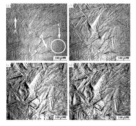

Figure 1. Bainite morphology evolution transformed at 330 °C for (a) 1 min, (b) 3 min, (c) 15 min, (d) 60 min. The samples were austenized at 1100 °C for 30 min.

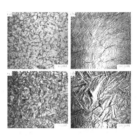

Figure 2. Comparison of bainite transformation for different times at 330 °C following austenization at 1000 and 1100 °C for 30 min: (a) 1 min, austenized at 1000 °C; (b) 1 min, 1100 °C; (c) 15 min, 1000 °C; (d) 15 min, 1100 °C.

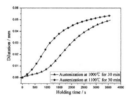

Figure 3. Dilation curves recorded by dilatometry in a thermal simulator during bainitic transformation at 330 °C.

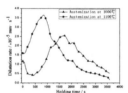

Figure 4. The effect of austenization temperature on the bainitic transformation rate.

and grow in the same direction parallel to each other towards the interior of the austenite grain [12], which is typically shown by circular area C in Figure 1a. The bainitic ferrites increase rapidly by the sympathetic nucleation, and finally cause impingement between the bainitic sheaves [13,14]. Figure 1 presents in situ observations of the morphological development of bainitic ferrite during further isothermal holding. The successive and repetitive nucleation and growth of sheaves of bainitic ferrite occur, and blocks and packets gradually appear within the austenite grain as shown in Figure 1. The nucleation and growth of bainite sheaves can be continuously observed using in situ measurement.

A comparison of bainitic transformation following austenization at 1000 and 1100 °C is shown in Figure 2. It can be observed that the bainite sheaves in the sample austenized at 1000 °C are obviously shorter than those at 1100 °C, showing that bainite morphology depends upon the parent austenite grains. Moreover, more bainite forms for the same transformation time in the sample austenized at 1100 °C than in the sample austenized at 1000 °C, confirming that a higher austenization temperature promotes the bainite transformation.

Table 1. Quantitative analysis of bainite transformation.

Time (s)		600	1200	1800	2400	3000	3600
Radial dilatation (mm)	1000 °C	0.0037	0.0112	0.0257	0.0376	0.0451	0.0497
	1100 °C	0.0132	0.0318	0.0423	0.0482	0.0514	0.0536
TransformationFraction (%)	1000 °C	7.53	22.52	51.68	75.75	90.86	100.00
	1100 °C	24.72	59.35	79.05	89.89	95.93	100.00

As previously mentioned, the bainite transformation can be continuously analyzed by in situ measurement. It is difficult, however, to quantitatively analyze the bainite transformation using in situ observation. In order to quantitatively analyze bainitic transformation, hot simulation experiments were conducted on a Gleeble 1500 simulator using the same treatment protocols as those used for LSCM. The dilatometric curves during the bainitic transformation process were recorded by dilatometry in a thermal-mechanical simulator. Figure 3 shows the dilatometric curves indicating the amount of bainitic transformation during isothermal holding at 330 °C following austenization at 1000 and 1100 °C. As mentioned above, the Bs, Bf and Ms are 423, 305 and 256 °C, respectively. This means that only bainitic transformation occurs during isothermal treatment at 330 °C. In other words, the dilatometric curves in Figure 3 represent the amount of bainite transformation in the tested steel. The ordinate in Figure 3 is the dilatometric amount of samples caused by bainitic transformation. The sample diameter expands with the phase transformation from γ-austenite to α-ferrite because of the lattice change from face-centered cubic (fcc) structure to body-centered cubic (bcc) structure. The densities of fcc and bcc are 0.74 and 0.68, respectively, leading to an increase in volume during bainitic transformation. The temperature of the sample changes from 329.6 to 329.7 °C during phase transformation, which demonstrates an almost stable temperature on the samples and a uniform temperature distribution from surface to interior. As mentioned above, the dimensions of Gleeble and LSCM samples are 8 mm diameter × 20 mm and 6 mm diameter × 4 mm, respectively. Therefore, the temperature distribution in the LSCM samples should also be uniform. This means that heat produced by the phase change has no effect on the dilation of the sample and the bainitic transformation on surface and interior is not affected by the temperature distribution in the samples. It can be observed from Figure 3 that bainitic transformation is greatly affected by austenizing temperature and the specimen austenized at 1100 °C presents a faster transformation kinetics, which means that phase transformation is accelerated by higher austenizing temperatures. On the other hand, the dilatometric curves reach about the same level after longer transformation time irrespective of the austenization conditions, and the total phase change amount is almost the same for the two austenization temperatures, which indicates that the austenization temperature has no significant influence on the overall amount of bainitic transformation after 60 min isothermal holding.

The changes in bainitic transformation rate with isothermal holding time are presented in Figure 4, indicating different maximum transformation rates for the two austenization temperatures. The maximum rate occurring at 848 s is 3.65×10^{-5} mm s^{-1} for the sample austenized at 1100 °C, and 2.55×10^{-5} mm s^{-1} occurring at 1696 s for the sample austenized at 1000 °C. The higher transformation rate at higher austenization temperature also proves that the transformation process has been accelerated. Higher austenization temperature results in the higher transformation rate, and the maximum rate also appears early with higher austenizing temperature, which corresponds to the phase transition rule in Figure 3.

Figure 2 shows that larger grains result from higher austenization, and Figure 3 demonstrates that higher austenization leads to faster transformation kinetics. This phenomenon is consistent with the result obtained by LSCM (Fig. 2); however, Godet et al. [12] reported contrary results, i.e. small austenite grains result in faster bainite transformation kinetics. They explained their results as follows. The reduction in grain size brings about an increase in the grain boundary area that accelerates the rate of transformation thanks to an enhanced nucleation rate. This influence has also been reported by Rees and Bhadeshia [14].

Austenite grain size after austenizing at 1100 °C is obviously larger than that at 1000 °C. The decrease in austenite grain size brings about an increase in the grain boundary area, leading to more nucleation sites. Therefore nucleation is faster in the sample with smaller austenite grain size. For growth, however, the situation is different. The driving forces for phase transformation of the tested steel are the same for the two processing routes except for austenization temperature because of the same undercooling. At the initial stage of bainite transformation, bainite sheaves in the large grain material grow more quickly under the same driving force due to fewer nuclei, which is proved by the growth rates in Figure 4. Not only is bainite sheaf growth rate in the initial stage of transformation in the sample austenized at 1100 °C larger, but the maximum growth rate also appears earlier (Fig. 4). As mentioned above, the total amount of bainitic transformation for the two austenization temperatures is about the same after 60 min isothermal treatment (Fig. 3). Only when the bainite growth rate with more nucleation sites in small-grained material is slower than that in large-grained material can essentially the same transformation amount be obtained. In the present study, experimental results suggest that growth plays a more important role than nucleation in bainite transformation.

It is assumed that the bainitic transformation finishes after 60 min isothermal treatment. The radial dilatation and phase transition fraction at different times for the two austenization temperatures are quantitatively analyzed and summarized in Table 1. It is apparent that the total diameter changes are almost the same for the two austenization temperatures, confirming that the total amount of bainitic transformation from γ-austenite to α-ferrite is hardly affected by the austenization temperature. However, the transformation rates are greatly accelerated at higher austenization temperatures. For example, the transformation fractions at 330 °C after 10 min isothermal holding are 7.53% and 24.72%, respectively, following austenization temperatures of 1000 and 1100 °C. The transformation fractions after 30 min holding are 51.68% and 79.05%, respectively, for the two austenization temperatures. The difference between the transformation fractions for the two temperatures decreases with the isothermal holding time, indicating that the transformation rate following higher austenization temperature decreases with holding time and the effect of austenization temperature on phase transition weakens with increasing transformation time.

It should be noted that bainitic transformation still takes place after 15 min (Fig. 3) and can be quantitatively analyzed in terms of the dilatometric data. However, phase transformation after 15 min cannot be analyzed by in situ micrography because it is hard to distinguish further bainite morphology evolution after 15 min from micrographs observed in situ as shown in Figure 1c and d.

Experimental results indicate that bainite transformation observed by LSCM is faster than that on a Gleeble simulator. LSCM micrographs show bainite transformation on the surface. The reason why surface transformation is more rapid is that the bainite reaction is displacive. These displacements cause strain when they occur in the interior of the sample, whereas the strain energy is relieved when the transformation happens at the free surface, making the bainite transformation in the proximity of a surface occur at a higher temperature [8].

So far a number of investigations on bainitic transformation using in situ observation have been reported as mentioned in the opening of this paper. Moreover, Singh and Bhadeshia studied the effects of ausforming on bainitic transformation using dilatometric data [15]. However, no report has been given on qualitative and quantitative investigations of bainite steel simultaneously. A new method, i.e. a combination of LSCM and dilatometer, is first proposed in this study to qualitatively analyze the morphology evolution of bainite ferrite by in situ observed micrographs and quantitatively investigate the transformation amount by dilatometric data simultaneously. The work in this study provides a new approach for the qualitative and quantitative analysis of bainitic phase transition.

In this paper, the nucleation and morphological evolution in a Fe–C–Mn–Si superbainite steel were directly observed by LSCM. The bainite ferrite nucleates at the grain boundary and in grains. The secondary bainite ferrite sympathetically nucleates on the preformed bainite sheaves. Bainite growth is characterized by the impingement of bainite sheaves, which results in an interlocked bainite microstructure. The bainitic transformation can be continuously monitored in real time by in situ measurement. The bainitic transformation, however, can only be qualitatively analyzed by in situ observation. Therefore, accurate dilatometry is used to quantitatively analyze the phase transition process. Qualitative and quantitative investigation of the phase transformation can be achieved using the new approach. This is the first time that transformations have been qualitatively and quantitatively investigated by a combination of LSCM and dilatometry. The work in present study provides a new approach for investigating phase transformations in steels.

The authors gratefully acknowledge financial support from the National Natural Science Foundation of China (NSFC) (No. 51274154), the National High Technology Research and Development Program of China (No. 2012AA03A504), the State Key Laboratory of Development and Application Technology of Automotive Steels (Baosteel Group) and a key project of the Hubei Education Committee (No. 20121101). G.X. is grateful to Dr. H.S. Hatem at McMaster University for valuable discussions.

Supplementary data associated with this article can be found, in the online version, at http://dx.doi.org/10.1016/j.scriptamat.2013.01.033.

[1] B.C. Muddle, J.F. Nie, Scr. Mater. 47 (2002) 187.
[2] C. Gupta, G.K. Dey, J.K. Chakravartty, D. Srivastav, S. Banerjee, Scr. Mater. 53 (2005) 559.
[3] M. Koo, P.G. Xu, Y. Tomota, H. Suzuki, Scr. Mater. 61 (2009) 797.
[4] D. Zhang, H. Terasaki, Y.I. Komizo, J. Alloys Compd. 484 (2009) 929.
[5] P. Kolmskog, A. Borgenstam, M. Hillert, P. Hedström, S.S. Babu, H. Terasaki, Y.I. Komizo, Metall. Mater. Trans. A 43 (2012) 4984.
[6] H. Yada, M. Enomoto, T. Sonoyama, ISIJ Int. 35 (1995) 976.
[7] M.K. Kang, M.X. Zhang, M. Zhu, Acta Mater. 54 (2006) 2121.
[8] J. Pak, D.W. Suh, H.K.D.H. Bhadeshia, Metall. Mater. Trans. A 43 (2012) 4520.
[9] M. Maalekian, R. Radis, M. Militzer, A. Moreau, W.J. Poole, Acta Mater. 60 (2012) 1015.
[10] D. Zhang, H. Terasaki, Y.I. Komizo, Acta Mater. 58 (2010) 1369.
[11] T. Ko, S.A. Cottrell, J. Iron Steel Inst. 172 (1952) 307.
[12] S. Godet, P. Harlet, F. Delannay, P.J. Jacques, Mater. Sci. Forum 426–432 (2003) 1433.
[13] X.L. Wan, H.H. Wang, L. Cheng, K.M. Wu, Mater. Charact. 67 (2012) 41.
[14] G.I. Rees, H.K.D.H. Bhadeshia, Mater. Sci. Technol. 8 (1992) 985.
[15] S.B. Singh, H.K.D.H. Bhadeshia, Mater. Sci. Technol. 12 (1996) 610.

Appendix 2: Academic Paper II

Materials and Design 31 (2010) 2891–2896

Contents lists available at ScienceDirect

Materials and Design

journal homepage: www.elsevier.com/locate/matdes

The development of Ti-alloyed high strength microalloy steel

Guang Xu *, Xiaolong Gan, Guojun Ma, Feng Luo, Hang Zou

Key Laboratory for Ferrous Metallurgy and Resources Utilization of Ministry of Education, Wuhan University of Science and Technology, Wuhan 430081, China

ARTICLE INFO

Article history:
Received 3 July 2009
Accepted 17 December 2009
Available online 24 December 2009

Keywords:
A. Ferrous metals and alloys
F. Microstructure
E. Mechanical
Ti-alloyed
Precipitates

ABSTRACT

Microalloy steels are generally Nb, V, or Nb, V, Ti composite microalloyed. Because of the high price of niobium and vanadium, the development of titanium microalloyed steels is a very interesting subject. In this study, steels with different Ti contents were refined and forged. Tensile tests were conducted and microstructures of samples were analyzed. Fine precipitates were observed using a transmission electron microscope. The results show that nanoscale Ti precipitates is the main factor enhancing strength of steels. The strength of steels increases with the Ti content. The optimum content range of titanium is between 0.04 and 0.10 wt.% while below 0.04 wt.% and higher than 0.10 wt.%, Ti has little effect on the strength of steels.

© 2010 Elsevier Ltd. All rights reserved.

1. Introduction

Microalloyed steels have been developed for many years and are widely used in industry today. The microalloyed steels are either Nb-alloyed, V-alloyed or Nb, V and Ti composite alloyed [1–3]. Steel alloyed with titanium alone is sparsely found. Nb, V and Ti microalloying elements can be found in steel as solution atoms or separate out as precipitates to retard austenite recrystallization and grain growth. By this way a fine grain microstructure can be obtained to enhance the strength of steels. A more important effect of microalloy elements on steels is their carbide and nitride precipitates. These precipitates can greatly improve the mechanical properties of steels.

The shortcoming of present microalloy steels is their high production cost because of the high price of microalloy elements. In the current situation of an economic recession in the iron and steel industry, this shortcoming is more obvious. Therefore, the development of titanium alloyed steel is interesting because of the low price of titanium. At present, although there are a few researches on Ti-alloyed hot strips produced by compact strip production (CSP) line [4,5], much work is needed for the development of Ti-alloyed steels. In order to develop high strength rebar, steels with different Ti contents were refined. Tensile tests were conducted and microstructures of samples were analyzed. Fine precipitates were observed using transmission electron microscope (TEM). The research results are summarized in this paper.

2. Experimental scheme

Using the literature [6–10] as a reference, the addition of Ti in steels was selected to be from about 0.04 wt.% to 0.1 wt.%. The chemical compositions of four different steels with varying Ti contents are given in Table 1. The chemical compositions of the steels were refined in a vacuum reduction furnace. Then refined steel was cast into round ingots with diameter of 60 mm. In order to obtain normal microstructure and precipitates, the ingots were heated to 1150 °C and kept for 30 min before forging. The ingots were forged to 40 mm in diameter followed by being heated to 1150 °C for 20 min. Then ingots were forged into round bar of 30 mm diameter. By above processing technology, precipitates can uniformly distributed in steels. The bars were machined to standard tensile test samples and tensile tests were conducted on a universal materials tester. The microstructures and inclusions of these three steels were analyzed using a scanning electron microscope (SEM). Precipitates were observed using TEM.

3. Results

3.1. Tensile tests

The standard tensile test samples were machined from forged steels. Tensile tests were conducted on a WAN-10000 materials testing machine. The engineering strain–stress curves are given in Fig. 1 and the mechanical properties of three steels in Table 1 are listed in Table 2.

---

* Corresponding author. Address: Mail box 131, School of Materials and Metallurgy, Wuhan University of Science and Technology, Wuhan 430081, China. Tel.: +86 027 63212211; fax: +86 027 86560679.
E-mail addresses: xuguang@wust.edu.cn, random88@163.com (G. Xu), GXL1028@126.com (X. Gan), gma@wust.edu.cn (G. Ma), wust_luofeng@163.com (F. Luo), zou_hang_happy@126.com (H. Zou).

0261-3069/$ - see front matter © 2010 Elsevier Ltd. All rights reserved.
doi:10.1016/j.matdes.2009.12.032

Table 1
Chemical compositions of three steels refined in this study (wt.%).

No.	C	Si	Mn	P	S	Ti
0#	0.22	0.346	1.34	0.025	0.023	0.0023
1#	0.24	0.338	1.30	0.024	0.018	0.034
2#	0.21	0.369	1.40	0.024	0.019	0.064
3#	0.25	0.369	1.38	0.024	0.021	0.108

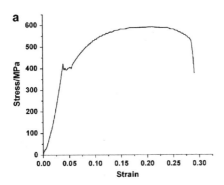
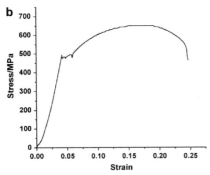
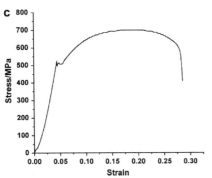

Fig. 1. The strain–stress curves of (a) steel 1; (b) steel 2; and (c) steel 3.

It can be seen from Table 2 that the yield strength and tensile strength increased with the increase of Ti content in steels. The basic chemical elements are similar for steels 1, 2 and 3 except for the Ti content. The yield strength and tensile strength of steels increased 80 MPa and 70 MPa respectively when the Ti content increased from 0.034% to 0.064%, while they increased 25 MPa and 20 MPa respectively when the Ti content increased from 0.064% to 0.108%. Therefore, the increase of strength is obvious when the Ti content is increased in the range of 0.04–0.10 wt.%.

3.2. Microstructure

Microstructures of steels were observed using a Zeiss optical microscope and Fig. 2 shows the microstructures of three steels. It is clear that microstructures of three steels are typical pearlite and ferrite, and their microstructures are similar.

3.3. Inclusions

The inclusions found in the microstructures of three steels were analyzed using the Philips XL30 SEM and the inclusion distribution, morphology and type (energy spectrum) are shown in Figs. 3–5.

Using the data obtained from energy dispersive X-ray spectrometer (EDS), it can be seen that the inclusions found in the steels are FeS, MnS, etc. In addition, very few Ti inclusions, $FeO \cdot TiO_2$, were also found in samples. Therefore, the inclusion compositions do not vary between the samples.

3.4. TEM analysis

TEM studies of three steels were conducted using JEM-2100FEF transmission electron microscope and the precipitates and particle types for the three steels are given respectively in Figs. 6–8. It can be seen from TEM analysis that (1) Sample 1, containing 0.034 wt.% Ti, has fewer precipitates than other two samples and the precipitated particles are globular Ti (C, N) whose size is about 30–150 nm while a few particles are less than 30 nm or larger than 150 nm; (2) Sample 2, containing 0.064 wt.% Ti, has more precipitates than sample 1 and has precipitated particles that are uniformly distributed, globular or irregular Ti (C, N) as well as a few Ti (S, C), and whose size is about 30–100 nm. A few particles are less than 30 nm or between 100 nm and 150 nm; and (3) Sample 3, containing 0.108 wt.% Ti, has the most precipitates among three samples and main precipitated particles are uniformly distributed globular Ti (C, N) whose size is about 30–100 nm while a few particles are between 10 nm and 30 nm or between 100 nm and 150 nm.

4. Discussion

It is well known that the most popular microalloy additions in steels are niobium, vanadium and titanium. As of today, the microalloyed steels currently in production are almost all Nb, V or Nb–V–Ti composite additions. Many researches on microalloyed steels with Nb and V additions have been published. For example, Gündüz and Cochrane [11] discussed the influence of cooling rate and tempering on precipitates of V-microalloyed steel and Shanmugama et al. [12] studied the microstructure of high strength niobium-containing pipeline steel. The microstructures and properties of a V-bearing alloy were investigated by Rasouli et al. [13]. But few research results on Ti-microalloyed steel have been given now. Microalloyed steel with Ti as main addition is a potential high strength alloy. The raw material of steels used in this study is a 20MnSi casting billet taken from a steel company. Differing amount of the Ti alloying element, 0, 0.034, 0.064, and 0.108 wt.% respectively, was added into the raw material to be refined to produce steels 0–3. The chemical compositions of four steels are similar except Ti content, whose compositions are given in Table 1. The yield strength for four steels is illustrated in Figs. 9 and 10 shows the relationship between uniform elongation and Ti content. From Figs. 9 and 10, we can see that the strength of steels

Table 2
Mechanical properties of three steels in this study.

No.	Uniform elongation (%)	Total elongation (%)	AR (%)	Upper yield stress (MPa)	Lower yield stress (MPa)	UTM (MPa)
0#	15.8	27.8	75	400	375	582
1#	16.80	23.68	47.76	420	395	591
2#	15.30	22.50	52.29	485	475	677
3#	13.70	23.70	45.10	525	500	698

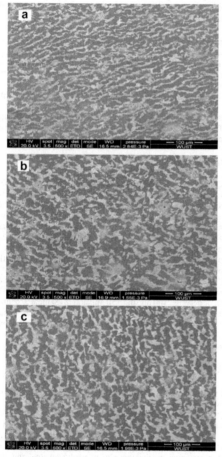

Fig. 2. Microstructures of (a) steel 1; (b) steel 2; and (c) steel 3.

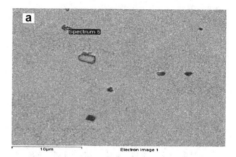

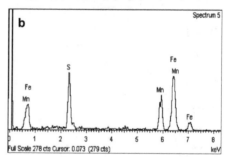

Fig. 3. Inclusion distribution and energy spectrum of steel 1.

increases with the increase of Ti content, while the elongation decreases with the increase of Ti content. The strength of steels has a sharp increase when Ti content changes between 0.04 and 0.07 wt.%.

The strength of steels depends on the strength of the matrix, the amount, size and distribution of precipitates etc. According to Hall–Petch relationship, the strength of steels can be expressed by:

$$\sigma_y = \sigma_i + \sigma_s + \sigma_p + \sigma_d + \sigma_g \qquad (1)$$

where σ_i is crystal lattice strengthening, σ_s is solution strengthening, σ_d is dislocation strengthening, σ_g is grain refining strengthening and σ_p is precipitation strength. From the analysis of the microstructure and inclusions for the four steels tested in this study, it has been shown that the microstructures and inclusions are similar, which means that the differences between the microstructures and inclusions for the four steels have little influence on mechanical properties. In addition, the basic compositions of four steels are almost identical, so their solution strengthening should be similar. Furthermore, the dislocation strengthening for the four steels should also be the same because the processing technology for these steels is the same after refining. Therefore, the strengthening of steels is caused by fine precipitates in steels.

According to the thermodynamic analyses in Ref. [14], the main forms of titanium precipitate are TiN, $Ti_4C_2S_2$ and TiC or Ti(C, N) composite particles, among which TiCs have the most important effect on steel strength because of their smaller size. The precipitation temperatures for TiN, $Ti_4C_2S_2$ and TiC particles are above 1500 °C, 1200 °C and 1000 °C, respectively. TiN and $Ti_4C_2S_2$ particles precipitate at higher temperature and easily grow to larger particles. From the SEM and TEM analyses in this paper, we found that precipitations of TiN and $Ti_4C_2S_2$ are larger than TiC particles, which is consistent with the results given by Zhou et al. [4] and Soto et al. [14]. In addition, TiN particles are coarse and cuboid in shape, while TiC particles are fine and more or less spherical (Figs. 3 and 8). Ooi and Fourlaris [15] also gave the similar results.

As aforementioned, steel 3, containing 0.108 wt.% Ti, has more precipitates than other steels. These precipitates, most of which

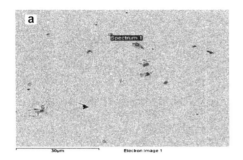

Fig. 4. Inclusion distribution and energy spectrum of steel 2.

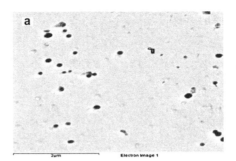

Fig. 6. Precipitates of steel and energy spectrum.

Fig. 5. Inclusion distribution and energy spectrum of steel 3.

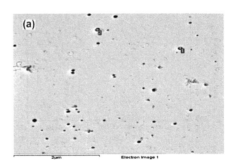

Fig. 7. Precipitates of steel 2 and energy spectrum.

are between 30 nm and 100 nm, are globular Ti (C, N) particles and uniformly distributed in steel. These nanoscale particles impede the dislocation movement during sample deformation to enhance the strength of the steel. On the other hand, steel 1, containing only 0.034 wt.% Ti, separated out fewer particles than steel 2 and steel 3. Corresponding strength is lower than other two Ti-alloyed steels.

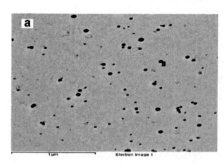

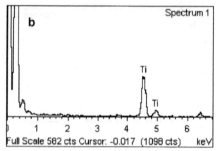

Fig. 8. Precipitates of steel 3 and energy spectrum.

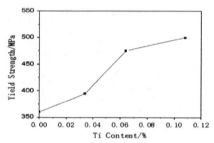

Fig. 9. The relationship between yield strength and Ti content.

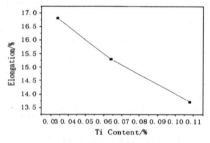

Fig. 10. Uniform elongation depending on Ti content.

Therefore, we can get the conclusion from these results that the main function of titanium in experimental steels is precipitation strength. Ti-alloyed steels can be produced to substitute the Nb-alloyed or Nb and V alloyed steels.

Table 3
Calculation results of Ashby–Orowan equation.

No.	X (nm)	f_v (%)	$\Delta\sigma_p$ (MPa)
1#	112.8	1.28	59.23
2#	32.94	0.72	117.88
3#	36.12	1.34	149.67

According to Gladmen [16], precipitate strengthening can be calculated by Ashby–Orowan equation:

$$\Delta\sigma_p = \frac{0.538 \cdot Gb \cdot f_v^{1/2}}{X} \ln\left(\frac{X}{2b}\right) \quad (2)$$

where $\Delta\sigma_p$ is increment of yield strength in MPa, G is shear modulus in MPa, 81,600 MPa for steels, b is Burger's vector in mm, 0.248 nm for ferrite, f_v is volume fraction of precipitate and X is diameter of precipitate particles in mm. The yield strength increments due to precipitate strengthening for samples 1–3 were calculated by above formula and the results are given in Table 3. The area fraction of precipitate is used here, which there is no much effect on calculation results. We can see from Table 3 that the strength increments by precipitate are 59.23 MPa, 117.88 MPa and 149.67 MPa, respectively. This calculation results are very close to the experimental results (Figs. 1 and 9).

5. Conclusions

Four different Ti-alloyed steels were refined and forged. Tensile tests, microstructure and inclusion analysis were conducted. The titanium carbonitrides precipitates were observed using TEM. Through these works, the following conclusions can be obtained:

(1) The strength of Ti-alloyed steels depends on the content of titanium. The obvious influence range of Ti on strength is between 0.04 wt.% and 0.07 wt.%.
(2) The differences in the observed microstructures and inclusions between the samples are negligible; microstructure plays little role in determining mechanical properties.
(3) The contribution of Ti content to strength can be attributed to precipitate strengthening.
(4) Ti-alloyed steels are a potential high strength steel. This study shows that Ti-alloyed steels can be used to substitute the Nb or Nb and V alloyed steels.

Acknowledgements

The authors would like to thank the Laiwu Iron and Steel (Group) Company for financially supporting this project. Also the English proofreading of paper by Mr. S. Crawford is greatly appreciated.

References

[1] Fu J, Wang ZB, Kang YL. Research and development of HSLC steels produced by EAF-CSP technology. In: Su TS, Li WX, editors. TSCR 2002:2002 international symposium on thin slab casting and rolling, 2002 December 3–5, Guangzhou, China. The Chinese Society for Metals; 2002. p. 301–4.
[2] Liu DL, Fu J, Kang YL. Oxide and sulfide dispersive precipitation and effects on microstructure and properties of low carbon steels. J Mater Sci Technol 2002;18(1):7–9.
[3] Mao XP. Microalloying technology on thin slab casting and direct rolling process. Beijing: Metallurgical Industry Press; 2008.
[4] Zhou J, Kang YL, Mao XP. Precipitation characteristic of high strength steels microalloyed with titanium produced by compact strip production. J Univ Sci Technol Beijing 2008;15(4):389–95.
[5] Bai MZ, Liu DL, Lou YZ, Mao XP, Li LJ, Huo XD. Effects of Ti addition on low carbon hot strips produced by CSP process. J Univ Sci Technol Beijing 2006;13(3):230–4.

[6] Fu J, Wu HJ, Liu YC. Nano-scaled iron–carbon precipitates in HSLC and HSLA steels. Sci China Ser E 2007;50(2):166–76.
[7] Garcia CI, Tokarz C, Graham C. Niobium HSLA steels producing the thin slab casting process: hot strip mill products, properties and applications. In: Su TS, Li WX, editors. TSCR 2002:2002 international symposium on thin slab casting and rolling, 2002 December 3–5, Guangzhou, China. The Chinese Society for Metals; 2002. p. 194–8.
[8] Yong QL. Microalloyed steels – Physical and mechanical metallurgy. Beijing: China Machine Press; 1989.
[9] Li Y, Wilson JA, Crowther DN. The effects of vanadium (Nb, Ti) on the microstructure and mechanical properties of thin slab cast steels. In: Su TS, Li WX, editors. TSCR 2002:2002 international symposium on thin slab casting and rolling, 2002 December 3–5, Guangzhou, China. The Chinese Society for Metals; 2002. p. 218–34.
[10] Kang YL, Yu H, Fu J. Morphology and precipitation kinetics of AlN in hot strip of low carbon steel produced by compact strip production. Mater Sci Eng A 2003;351(1–2):265–71.
[11] Gündüz S, Cochrane RC. Influence of cooling rate and tempering on precipitation and hardness of vanadium microalloyed steel. Mater Des 2005;26(6):486–92.
[12] Shanmugama S, Misraa RDK, Hartmannb J, Jansto SG. Microstructure of high strength niobium-containing pipeline steel. Mater Sci Eng A 2006;441(1–2):215–29.
[13] Rasouli D, Khameneh AS, Akbarzadeh A, Daneshi GH. Optimization of mechanical properties of a micro alloyed steel. Mater Des 2009;30(6):2167–72.
[14] Soto R, Saikaly W, Bano X, Issartel C, Rigaut G, Charai A. Statistical and theoretical analysis of precipitates in dual-phase steels microalloyed with titanium and their effect on mechanical properties. Acta Mater 1999;47(12):3475–81.
[15] Ooi SW, Fourlaris G. A comparative study of precipitation effects in Ti only and Ti–V Ultra Low Carbon (ULC) strip steels. Mater Charact 2006;56(3):214–26.
[16] Gladman T. Precipitation hardening in metals. Mater Sci Technol Ser 1999;15(1):30–6.

Appendix 3: Academic Paper III

Materials and Design 84 (2015) 95–99

Contents lists available at ScienceDirect

Materials and Design

journal homepage: www.elsevier.com/locate/jmad

The effects of Nb and Mo addition on transformation and properties in low carbon bainitic steels

Haijiang Hu [a,b], Guang Xu [a,*], Li Wang [b], Zhengliang Xue [a], Yulong Zhang [b], Guanghui Liu [c]

[a] *The State Key Laboratory of Refractories and Metallurgy, Hubei Collaborative Innovation Center for Advanced Steels, Wuhan University of Science and Technology, Wuhan 430081, China*
[b] *State Key Laboratory of Development and Application Technology of Automotive Steels (Baosteel Group), Shanghai 201900, China*
[c] *Daye Special Steel Company Limited, Hubei Xinyegang Steel Company Limited, Huangshi 435001, China*

ARTICLE INFO

Article history:
Received 24 October 2014
Received in revised form 17 June 2015
Accepted 23 June 2015
Available online 30 June 2015

Keywords:
Bainitic steel
Bainitic transformation
Microstructure
Property

ABSTRACT

Four low carbon bainite steels were designed to investigate the effects of Mo and Nb addition on bainitic transformation, microstructures and properties by metallographic method and dilatometry. The results show that single Nb addition retards bainitic transformation in low carbon bainite steels, although it can improve strength by refining microstructures. Moreover, Mo addition is effective to improve the strength of low carbon bainite steel by promoting bainitic transformation and single Mo addition has a better strengthening effect than single Nb addition. Further, in Mo bearing steel, Nb addition refines bainite sheaves, but meanwhile hinders bainitic transformation because of smaller austenite grains. Consequently, the composite strengthening effect of Mo and Nb addition has little improvement compared with individual addition of Mo in low carbon bainite steels.

© 2015 Elsevier Ltd. All rights reserved.

1. Introduction

Low carbon bainitic steels are commonly designed with the addition of niobium (Nb) and molybdenum (Mo) to achieve favorable combination of strength and toughness. The microalloying element Nb is known to improve microstructure and properties by precipitation strengthening and grain refinement [1–3]. Molybdenum addition can separate the bainitic transformation zone to obtain desired bainitic microstructure over a wide range of cooling rates [4,5]. Some investigations have been done on the exploration of the effects of Nb and Mo on the phase transition, microstructures and properties of various steels. Wang et al. [6] studied the influence of Nb on microstructure and property of low-carbon Mn-series air-cooled bainitic steel. They claimed that the amount of bainite increases by Nb addition and bainite clusters are finer in Nb-bearing steel. Hausmann et al. [7] investigated the effects of Nb on transformation behavior and mechanical properties of bainitic–ferritic steels. Niobium addition was found to indirectly accelerate upper bainite formation through grain refinement. However, Chen et al. [8] studied the effect of Nb on the formation of bainite ferrite in a low carbon HSLA steel and concluded that precipitates of Nb(C,N) retard formation of bainite ferrite during continuous cooling process. Therefore, the effect of Nb on bainitic transformation in low carbon bainitic steel is worthwhile to further investigation.

* Corresponding author.
E-mail address: xuguang@wust.edu.cn (G. Xu).

As to element Mo, Khare et al. [9] investigated the influence of Mo on bainite transformation kinetics and found that the effect of Mo is not detectable with Mo amounts varying between 0% and 0.25% in a 0.32% C bainite steel. But Kong and Xie [10] showed that the bainite starting temperature (Bs) of a low carbon microalloyed steel was obviously reduced and the size of microstructure became smaller when 0.40 wt.% Mo was added to the steel. Chen et al. [11] claimed that it is necessary to add suitable Mo to improve the toughness and strength of high Nb-bearing X80 pipeline steels (0.26% Mo, 0.07% Nb), especially, the increase of the strength. They considered that Mo addition can suppress the polygonal ferrite (PF) transformation, and decrease the transformation temperature of bainite ferrite, resulting in refined transformed products, so the mechanical properties are improved. Moreover, Sourmail and Smanio [12] claimed that the effect of Mo on bainite transformation is not clear with a negligible or no retardation influence of bainite formation kinetics, although the calculated effect on the driving force leads to an expected acceleration. Thus the influence of Mo on bainite transformation requires more attention.

In addition, Zhu et al. [13] researched the influences of boron (B) and B + Nb on the bainitic transformation in low carbon steels, suggesting that the addition of B decreased only slightly the bainite transformation temperature at low cooling rates, whereas the combined addition of B + Nb decreased greatly the transformation temperature. It was also found that a high hardenability can be obtained due to a strong synergistic effect arising from the combined addition of B and Nb [14] or B and Mo [15].

http://dx.doi.org/10.1016/j.matdes.2015.06.133
0264-1275/© 2015 Elsevier Ltd. All rights reserved.

Summarizing existing literature, it indicates that some researchers have reported the effects of individual Nb or Mo addition on microstructure and property in low carbon bainitic steels. However, so far few studies have been conducted on the effects of combined addition of Nb and Mo on the transformation, microstructure and property in low carbon bainitic steels. Four kinds of low carbon bainitic steels have been designed in this study to investigate the effects of Nb and Mo on bainitic transformation, microstructure and properties. Heat treatment experiments with the same procedure were performed on ThermecMaster-Z hot thermal–mechanical simulator followed by microstructure and property analyses as well as quantitative characterization of bainite transformation with dilation data. The purpose of the present study is to clarify the influences of combined addition of Nb and Mo on the transformation, microstructure and properties in low carbon bainitic steels. The results will be helpful for optimizing the composition design of Nb–Mo alloying low carbon steels.

2. Experimental procedure

Four low carbon bainitic steels with different chemical compositions were refined using laboratory-scale vacuum furnace. Cast ingots were hot-rolled to 12 mm thick plates. As shown in Table 1, single Nb or Mo addition was to study the effects of individual element on transformation and properties in the low carbon bainitic steel, and the combined addition of Nb and Mo was considered to investigate the composite influence.

Samples for dilatometric study were machined to a cylinder of 6 mm diameter and 12 mm height. The top and bottom surfaces of samples were polished conventionally to keep the measurement face level. The experiments were conducted according to the procedure shown in Fig. 1 on a ThermecMaster-Z hot thermal–mechanical simulator equipped with a LED dilatometer to quantitatively analyze the bainitic transformation of four steels. The specimens were heated to 1000 °C at a rate of 10 °C/s and held for 900 s to achieve a homogeneous austenitic microstructure, followed by cooling to 350 °C at a rate of 10 °C/s. For these four low-carbon bainite steels, ferrite or pearlite transformation can be avoided by the cooling rate of 10 °C/s. After isothermal holding for 3600 s at 350 °C for bainite precipitation, the samples were air cooled to room temperature. Additionally, in order to investigate the properties of the tested steels, 140 × 20 × 10 mm blocks were cut from hot-rolled sheets and heat treated using the same procedure shown in Fig. 1.

The specimens were mechanically polished and etched with a 4% nital solution for microstructure examination by scanning electron microscopy (SEM). Both the bainite sheaves and grain morphology were examined using a Nova 400 Nano field emission scanning electron microscope operated at an accelerating voltage of 20 kV. Grain sizes of specimens were calculated by Image-Pro Plus software, and tensile tests were carried out on UTM-5305 electronic universal tensile tester at room temperature. Tensile specimens were prepared according to ASTM standard and the strain rate is ~4 × 10^{-3}/s.

3. Results

3.1. Microstructures

Fig. 2 shows typical SEM microstructures of four steels after isothermal holding for 3600 s at 350 °C following austenization at 1000 °C for

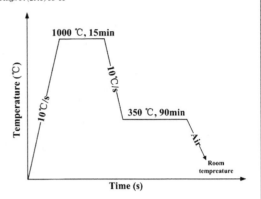

Fig. 1. Experimental procedure.

900 s. In order to clearly observe the amount of bainite in steels with different alloying elements, bainite was colored orange. For low carbon bainitic steels, the classification method proposed in Ref. [16] was used in the present work to identify the microstructure, i.e. the microstructure is classified as PF, granular bainite (GB), bainite ferrite (BF) and martensite (M). It can be observed clearly that the microstructures of four specimens mainly consist of lath-like BF and prior-austenite grain boundaries are well defined, as shown by arrows in Fig. 2a. Without Mo addition, some PF was formed in steel A (base steel) and steel C (Nb steel), as marked by arrow in Fig. 2c. According to the micrographs, the volume fraction of bainite was calculated by Image Pro-Plus software. The volume fractions of bainite of steels A, B, C, and D are 55.8%, 68.4%, 34.3% and 47.8%, respectively. It reveals that the sample with single Mo addition has the largest amount of bainite ferrite, while the one with only Nb addition has the smallest percentage of bainite. Comparison of the bainitic microstructures of four steels indicates that more bainite sheaves can be obtained with single addition of Mo (Fig. 2b). However, it should be pointed out that although single addition of Mo obviously promotes bainitic transformation, bainite fraction reduces after composite addition of Mo and Nb (Fig. 2b and d).

The original austenite grain size of tested samples, which influences the bainite morphology [17,18], was calculated by Image-Pro Plus software. The measurement was based on the total morphology of prior austenite grains shown in ellipse area in Fig. 2d. The average value of prior-austenite grain size was obtained with multiple measurements in Table 2. The results indicate that smaller austenite grains were obtained after austenization at 1000 °C with the addition of Nb. Compared to the average grain size of 39.4 μm in steel A (base steel), the grain sizes of steel B (Nb steel) and steel D (Mo + Nb steel) were reduced by ~33.2% and ~62.2%, respectively. It means that the combined addition of Mo and Nb has a better refining effect than single Nb addition.

3.2. Mechanical properties

The engineering stress–strain curves of four steels with different composition grades are shown in Fig. 3, and the corresponding mechanical properties are given in Table 3. The yield and tensile strength are improved by single addition of Mo or Nb and combined addition of Mo and Nb, while total elongations (TE) for four steels exhibit no significant change. Compared with steel A (base steel), the yield strength (YS) and ultimate tensile strength (UTS) of Mo addition steel (steel B) increase by 439 MPa and 345 MPa, while the YS and UTS increments of Nb additional steel (steel C) are only 152 MPa and 102 MPa, respectively. It is apparent that the increase of YS and UTS for Nb additional steel results from the refinement of grains (Fig. 2c), while the improvement of strength for Mo addition steel is due to more bainite in microstructure (Fig. 2b).

Table 1
Chemical compositions of steels (wt.%).

Steel	C	Si	Mn	Nb	Mo
A (base)	0.215	1.535	2.013	/	/
B (Mo)	0.221	1.504	1.976	/	0.138
C (Nb)	0.218	1.497	2.034	0.025	/
D (Mo + Nb)	0.214	1.499	2.023	0.025	0.142

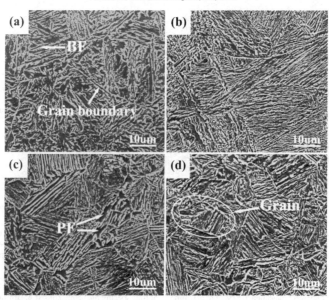

Fig. 2. SEM microstructures of four low carbon bainite steels after isothermal holding for 3600 s: (a) base, (b) Mo addition, (c) Nb addition, and (d) Mo and Nb addition.

Although the decreased average size of prior austenite grain leads to strength increasement for single Nb addition steel, the single Mo addition plays more significant effect on strength improvement for low carbon bainite steels due to the strengthening effect of more bainite fraction (Fig. 2b). Additionally, it is very interesting to note that the strength and elongation of steel B (single Mo) and steel D (Mo + Nb) have no obvious difference, suggesting that no further strength improvement occurs by Nb addition in Mo bearing steel compared to individual Mo addition steel. In other words, it seems that Nb should not be added in C–Mn–Si–Mo low carbon bainitic steels. With regard to the composition design standards of low carbon bainitic steels in industrial production, some researchers reported that Mo and Nb are added together [7,8]. This viewpoint is arguable.

3.3. Thermal dilatometry

The metallographical microstructures can be used to qualitatively analyze the effect of Mo and Nb on bainite microstructure. In order to quantitatively investigate the influence of combined Mo + Nb addition on bainite transformation, dilatometric experiments were conducted on the thermo-mechanical simulator. According to the recorded dilatometric data, dilatation curves of four steels were plotted. Fig. 4 shows dilatations as a function of holding time during thermal holding at 350 °C, where the beginning of isothermal holding was selected as the zero point of abscissa and ordinate axes. The transformation temperature was constant and no extra force was applied on the sample during isothermal holding, thus the dilatation in Fig. 4 represents the real bainite transformation amount. It can be observed that the final bainite fraction increases with the addition of Mo, but obviously reduces with the addition of Nb, meaning that Nb addition retards the bainitic transformation in low carbon bainite steels. The baintie transformation amount of steel D (Mo + Nb steel) is smaller than that of steel A (base steel), indicating that the addition of Nb in Mo bearing bainite steel has no positive effect on bainitic transformation. In summary, individual addition of Mo has a much better promoting effect on bainite reaction than combined Mo and Nb addition.

Fig. 5 shows the relationship of volume fractions of bainite with time during isothermal holding at 350 °C. It shows that the steels with Mo addition (steel B and steel D) completed bainite transformation prior to the steels without Mo (steel A and steel C). The result indicates that

Table 2
Prior-austenite grain sizes of four steels (μm).

	Base	Mo	Nb	Mo + Nb
Prior-austenite grain size	39.4 ± 9.1	40.8 ± 9.4	26.3 ± 7.3	14.9 ± 4.3

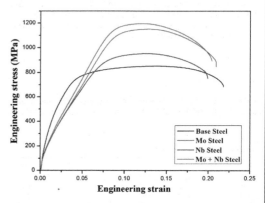

Fig. 3. Stress–strain curves of four steel with different compositions.

Table 3
Mechanical properties of samples with different compositions.

Steel	YS (MPa)	UTS (MPa)	TE (%)	UTS × TE (GPa%)
A (base)	617	853	21.8	18.6
B (Mo)	1056	1198	20.3	24.3
C (Nb)	769	955	19.9	19.0
D (Mo + Nb)	1009	1151	20.9	24.1

Mo accelerates bainite transformation. Meanwhile, compared to steel A, bainite transformation in steel D was slightly slower, suggesting that Nb obviously weakens the acceleration effect of Mo on bainite transformation.

4. Discussions

4.1. Influence of Mo

SEM micrographs (Fig. 2) show that steel A mainly consists of BF sheaves and small amount of PF and martensite, while no PF exists in steel B. As reported by some researches [19–21], bainitic transformation is characterized by incomplete reaction, thus the non-transformed austenite after isothermal holding could transform into martensite. The occurrence of ferrite in steel A indicates that high temperature phase transition happens during the cooling process. However, with the Mo addition, the ferrite transformation is avoided, resulting in more bainite fraction. It was reported that Mo causes a separation of the bainite C-curve and extends the bainite formation field [4,5]. Therefore, more undercooled austenite can transform into BF in the Mo addition steel.

Based on the quantitative analysis results in Fig. 4, the dilatations of steel A (base steel) and steel B (Mo steel) are 0.0236 mm and 0.0279 mm, respectively, demonstrating a 18.2% increment in bainite microstructure with a 0.138% Mo addition. The results in dilatometric test are consistent with the metallographical microstructures. The UTS of steel B (1198 MPa) increases by 40.4% compared with steel A (853 MPa). For bainitic steel, more fine bainite can improve mechanical properties. In the present work, the volume fraction of bainite in steel B increases by 12.6% compared with steel A, resulting in an increment of 345 MPa in strength. It demonstrates that a small amount of Mo addition in low carbon bainitic steel greatly improves strength of steel with total elongation almost unchanged. The addition of Mo promotes

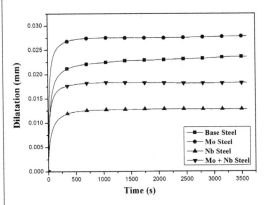

Fig. 4. Dilation curves recorded by dilatometer on thermal simulator during bainitic transformation at 350 °C.

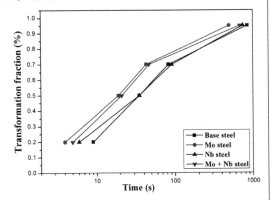

Fig. 5. Bainitic transformation volume fraction with time during isothermal holding at 350 °C.

bainitic formation and results in more bainite in microstructure, which plays an important role in the improvement of strength of steel B.

4.2. Influence of Nb

SEM micrographs indicate that steel A (base steel) and steel C (Nb steel) contain small amount of PF, showing that Nb addition nearly has little influence on the avoidance of ferrite transformation. However, comparing the prior-austenite grain size of two steels, it is clear that the austenite grains were markedly refined with 0.025% Nb addition, leading to higher strength. It is well known that Nb in solid solution state in austenite could inhibit the growth of austenite grains due to the strong drag effect on the migration of grain boundaries [22]. Thus the average austenite grain size prior to phase transition was 26.3 μm for the Nb steel, much smaller than that of steel A. The volume fraction of bainite in steel C reduces by 21.5% compared with steel A, which is negative for strength. However, the tensile strength of Nb steel still increases by 92 MPa compared with base steel in spite of lesser amount of bainite. The strength increment may be explained by grain refinement, sold solution strengthening of Nb atoms and precipitation strengthening of carbonitrides formed during the cooling process [23].

The single addition of Nb also affects bainite transformation kinetics, which has been reported by several authors. Both Meyer et al. [24] and Hoogendoorn and Spanraft [25] reported that Nb addition could promote the formation of more low temperature transformation product, such as granular bainite. On the other hand, it has been shown that increasing the niobium content in solid solution in a low carbon steel (0.14% C–1.6% Mn–0.4 Si) retards the bainite transformation upon cooling, but this effect is rather small [26]. Däcker et al. [27] revealed that the addition of 0.04% Nb in a 0.05% C–1.5% Mn–0.3% Si steel has no significant effect on bainite phase transformation kinetics. However, the result in this study is different. According to the dilation diagram, the total dilatation of steel C (Nb steel) is only 0.0128 mm, reduced by 45.8% compared to steel A (base steel, 0.0236 mm). It indicates that 0.025% Nb addition obviously hinders the bainite transformation. As mentioned previously, grain size is much smaller in steel C than that in steel A. It has been proved that small austenite grains retard bainitic transformation [17,28]. The decrease in austenite grain size brings about an increase in the grain boundary area, leading to more nucleation sites. The driving forces for phase transformation of tested steels are the same because of the same transformation temperature. Thus bainite sheaves in the large grain material grow more slowly under the same driving force due to more nuclei. In addition, small austenite grains bring more restrictions, making it difficult for bainite sheaves to grow. Chen et al. [8] studied the effect of Nb on the formation of bainite ferrite in a low carbon HSLA steel and obtained similar result. But they explained that precipitates of Nb(C,N) retard formation of bainite ferrite during continuous cooling process. In summary, the addition of Nb hinders bainitic formation although it improves the strength of low carbon bainitic steels by grain refinement.

4.3. Composite effect of Mo and Nb

The previous discussions show that single Mo addition can promote the bainite transformation, leading to phase transition strengthening, and single Nb addition can refine microstructures, resulting in grain refine strengthening. In order to investigate the effects of combined addition of Mo and Nb on bainitic transformation kinetics, microstructures and properties, low carbon bainite steel with Mo and Nb addition (steel D) was designed. It can be observed from SEM micrographs that steel D has a smaller volume fraction of bainite than steel B (Mo steel). Thus the Nb addition weakens the promotion function of Mo on bainite transformation. Compared with base steel, the average grain size in steel D is much smaller due to Nb addition, resulting in the 8.0% decrease of volume fraction of bainite. However, weakening of strength due to reduced bainite fraction is compensated by the refinement strengthening. The combined effect of Nb and Mo elements finally makes the strength of steel D increase by 298 MPa compared with steel A. Also from the dilation curves, the dilatation of Mo and Nb bearing steel is 0.0183 mm, decreased by 34.4% compared to the 0.0279 mm expansion of Mo bearing steel, which is consistent with metallographic results (Fig. 2). In addition, the product of tensile strength and total elongation is 24.1 GPa% for steel D and 24.3 GPa% for steel B, which indicates that no further improving effect on comprehensive property by combined Mo and Nb addition is obtained compared to single Mo addition. In Mo bearing steel, Nb addition refines bainite sheaves due to smaller austenite grains, but hinders the bainitic transformation because of smaller austenite grains. Therefore, the composite strengthening effect of Mo and Nb addition has no significant improvement than individual addition of Mo in low carbon bainite steels. Therefore, it seems that Nb addition should be avoided in Mo bearing low carbon bainite steel. The main novelty of the present work is that the combined effect of alloying elements Nb and Mo on bainite transformation, microstructure and properties is investigated. The Nb addition in Mo bearing steel cannot further improve the strength due to its negative effect for the promotion function of Mo on bainite transformation. The result provides a theoretical basis for the composition design of low carbon bainitic steel.

5. Conclusions

Four low carbon bainite steels were designed in the present work. Metallographic method and dilatometry were combined to investigate the effect of Mo and Nb addition on bainite transformation, microstructures and properties. Experimental results indicate that single Nb addition retards bainitic transformation in low carbon bainite steels, although it can improve strength by refining microstructures. Mo addition is effective to improve the strength of low carbon bainite steel by promoting bainitic transformation, and single Mo addition has a better strengthening effect than single Nb addition. In addition, the Nb addition weakens the promotion function of Mo on bainite transformation. In Mo bearing steel, Nb addition refines bainite sheaves due to smaller austenite grains, but hinders the bainitic transformation because of smaller austenite grains. Thus the composite strengthening effect of Mo and Nb addition has no significant improvement than individual addition of Mo in low carbon bainite steels. Therefore, Nb addition could be avoided in Mo bearing low carbon bainite steel.

Acknowledgments

The authors gratefully acknowledge the financial supports from the National Natural Science Foundation of China (NSFC) (no. 51274154), the National High Technology Research and Development Program of China (no. 2012AA03A504), the State Key Laboratory of Development and Application Technology of Automotive Steels (Baosteel Group) (BS2012-01) and the Key Project of Hubei Education Committee (no. 20121101).

References

[1] A.J. DeArdo, Int. Mater. Rev. 48 (2003) 371.
[2] D. Krizan, B.C. De Cooman, Steel Res. Int. 79 (2008) 513.
[3] S. Traint, A. Pichler, R. Sierlinger, H. Pauli, E. Werner, Steel Res. Int. 77 (2006) 641.
[4] V.F. Zackay, W.M. Justusson, Iron Steel Inst. (1962) 14.
[5] F.B. Pickering, Physical Metallurgy and the Design of Steels, Applied Science Publishers, Ltd., London, 1978.
[6] Y.W. Wang, C. Feng, F.Y. Xu, B.Z. Bai, H.S. Fang, J. Iron Steel Res. Int. 17 (2010) 49.
[7] K. Hausmann, D. Krizan, K. Spiradek-Hahn, A. Pichler, E. Werner, Mater. Sci. Eng. A 588 (2013) 142.
[8] Y. Chen, D.T. Zhang, Y.C. Liu, H.J. Li, D.K. Xu, Mater. Charact. 84 (2013) 232.
[9] S. Khare, K. Lee, H.K.D.H. Bhadeshia, Int. J. Mater. Res. 100 (2009) 1513.
[10] J.H. Kong, C.S. Xie, Mater. Des. 27 (2006) 1169.
[11] X.W. Chen, G.Y. Qiao, X.L. Han, X. Wang, F.R. Xiao, B. Liao, Mater. Des. 53 (2014) 888.
[12] T. Sourmail, V. Smanio, Acta Mater. 61 (2013) 2639.
[13] K.Y. Zhu, C. Oberbillig, C. Musik, D. Loison, T. Iung, Mater. Sci. Eng. A 528 (2011) 4222.
[14] T. Hara, H. Asahi, R. Uemori, H. Tamehiro, ISIJ Int. 44 (2004) 1431.
[15] H. Asahi, ISIJ Int. 42 (2002) 1150.
[16] F.R. Xiao, B. Liao, Y.Y. Shan, K. Yang, Mater. Charact. 54 (2005) 417.
[17] G. Xu, F. Liu, L. Wang, H.J. Hu, Scr. Mater. 68 (2013) 833.
[18] F. Liu, G. Xu, Y.L. Zhang, H.J. Hu, L.X. Zhou, Z.L. Xue, Int. J. Miner. Metall. Mater. 20 (2013) 1060.
[19] H.K.D.H. Bhadeshia, Bainite in Steels, 2nd ed. IOM Communications, London, 2001.
[20] X.L. Wang, K.M. Wu, F. Hu, L. Yu, X.L. Wan, Scr. Mater. 74 (2014) 56.
[21] F.G. Caballero, M.K. Miller, C. Garcia-Mateo, J. Cornide, J. Alloys Compd. 577S (2013) S626.
[22] B. Fu, W.Y. Yang, M.Y. Lu, Q. Feng, L.F. Li, Z.Q. Sun, Mater. Sci. Eng. A 536 (2012) 265.
[23] S. Hashimoto, S. Ikeda, K. Sugimoto, S. Miyake, ISIJ Int. 44 (2004) 1590.
[24] L. Meyer, F. Heisterkamp, W. Mueschenborn, Proc. Conf. Microalloying 75 (1977) 153.
[25] T.M. Hoogendoorn, M.J. Spanraft, Proc. Conf. Microalloying 75 (1977) 75.
[26] G.J. Rees, C. Perdrix, T. Maurickx, H.K.D.H. Bhadeshia, Mater. Sci. Eng. A 194 (1995) 179.
[27] C.-Å. Däcker, M. Green, J.-L. Collet, K. Zhu, N. Kwiaton, R. Kuziak, Z.I. Olano, C. Stallybrass, C. Luo, Bainitic Hardenability, RFSR, Contract No-2007-00023, 2010.
[28] F. Hu, P.D. Hodgson, K.M. Wu, Mater. Lett. 122 (2014) 240.

Appendix 4: Academic Paper IV

Materials Science & Engineering A 626 (2015) 34–40

Contents lists available at ScienceDirect

Materials Science & Engineering A

journal homepage: www.elsevier.com/locate/msea

New insights to the effects of ausforming on the bainitic transformation

Haijiang Hu [a], Hatem S. Zurob [b,*], Guang Xu [a], David Embury [b], Gary R. Purdy [b]

[a] *The State Key Laboratory of Refractories and Metallurgy, Hubei Collaborative Innovation Center for Advanced Steels, Wuhan University of Science and Technology, Wuhan 430081, China*
[b] *Department of Materials Science and Engineering, McMaster University, 1280 Main St. W., Hamilton, Canada L8S4L7*

ARTICLE INFO

Article history:
Received 9 September 2014
Received in revised form
11 December 2014
Accepted 13 December 2014
Available online 20 December 2014

Keywords:
Ausforming
Bainite transformation
Mechanical stabilization
Dilatometry

ABSTRACT

The effects of prior deformation conditions on the bainitic transformation are investigated using dilatometry and optical metallography. The influences of both deformation temperature and strain on bainite transformation kinetics and morphology are examined in superbainitic carbide-free steel. The transformation is retarded by deformation at high temperature due to the retardation of bainite growth by dislocation debris in deformed austenite. At low deformation temperatures, the initial transformation rate is accelerated due to the presence of prolific nucleation sites in deformed austenite. The growth rate is, again, retarded by deformation. The combined effect of accelerated nucleation and retarded growth results in a complex dependence of the phase transformation kinetics on applied strain. For small deformation strains at 300 °C, the overall transformation kinetics is faster than that in the non-deformed material. From a practical point of view, this provides an important opportunity to reduce the processing times for carbide-free bainitic steels. Interestingly, both the final transformed bainite and retained austenite fraction under these conditions are much higher than the non-deformed material. The retardation of growth dominates the overall transformation kinetics at large strains and low temperature leading to a high volume fraction of martensite in the final microstructure.

© 2014 Elsevier B.V. All rights reserved.

1. Introduction

Plastic deformation can strongly influence phase transformations in steels. It is generally accepted that ferrite and pearlite transformations are promoted by prior deformation [1,2], while the formation of martensite is retarded [3–5] by prior deformation. For bainite precipitation, however, the effect of ausforming on the transformation kinetics is not very clear. Shipway et al. [6] summarized the effects of ausforming on bainitic transformation and claimed that the motion of the transformation interface is hindered by the accumulated debris of dislocations in the deformed austenite. Larn et al. [7] studied the effect of compressive deformation at 800 °C on subsequent bainite precipitation in Fe–Mn–Si–C steels and found that the overall transformation kinetics became slower and the final attained amount of bainite decreased in deformed austenite. Chiou et al. [8] investigated the effect of prior compressive deformation of austenite on the toughness in an ultra-low carbon bainitic steel. They argued that compressive deformation stifled the formation of sheaf-like parallel plates of bainitic ferrite. Yang et al. [9] and Davenport [10] reported that the bainitic transformation was retarded by ausforming to large strain. Lee et al. [11] gave similar results in their study on the effect of plastic deformation on the formation of acicular ferrite and found that the transformation to acicular ferrite was hindered and the final fraction of acicular ferrite was reduced in plastically deformed austenite. The above investigations seem to suggest that ausforming not only decreases the transformation rate, but also reduces the final amount of bainite formed.

Other researchers claimed that deformation accelerates the initial transformation rate (nucleation) but retards the later stages which are dominated by growth. Bhadeshia [12] and Singh [13] found that although the deformation of austenite did introduce more heterogeneous nucleation sites, the growth of the bainitic plates was drastically reduced by deformation, resulting in slower overall transformation kinetics. The nucleation rate is larger in the heavily deformed regions but the overall rate of transformation is reduced because each nucleus then transforms to a smaller volume. Maki [14] claimed that although deformed austenite transformed faster initially, the net volume fraction of bainite that formed decreased when compared with undeformed austenite. Freiwillig et al. [15] also reported an initial acceleration of transformation to bainite from deformed austenite, but final transformed fraction decreased. Jin et al. [16] found that the

* Corresponding author.

http://dx.doi.org/10.1016/j.msea.2014.12.043
0921-5093/© 2014 Elsevier B.V. All rights reserved.

isothermal decomposition of deformed austenite was significantly promoted as the incubation period was remarkably shortened for small strain (e.g. 5%) but the final amount of bainite formed was reduced. Edwards and Kennon [17] reported similar results.

In few cases, it was reported that the whole transformation process is promoted by ausforming. Gong et al. [18,19] studied the effect of ausforming on kinetics, morphology and crystallography of nano-scale bainite steel. It was found that the bainite transformation at 300 °C was promoted after the steel was compressed by small reduction at this temperature while ausforming at high temperature had little influence on bainitic transformation. These authors did not investigate the effect of large deformation strains.

In summary, three different results have been reported, indicating that the influence of deformation on the bainite transformation is still a controversial topic. It is noteworthy that most of earlier work has not systematically considered the effect of deformation temperature and strain on the bainitic transformation. Therefore, in order to further investigate the effects of ausforming on the bainitic transformation both deformation temperature and strain are varied in the present work. The steel of interest is a simple, Fe–C–Mn–Si, carbide-free bainitic steel. The transformation kinetics is examined using metallography and dilatometric analysis.

2. Experimental

A superbainite steel containing 0.4C, 2.8 Mn, 2.0 Si (wt%) was used in the present study. The steel was cast in the form of a 40 kg ingot using laboratory-scale vacuum furnace at ArcelorMittal Dofasco (Hamilton, Canada). The material was then hot-forged and air-cooled to room temperature. The steel was then tempered at 650 °C for 15 h in order to facilitate machining.

The specimens for the thermomechanical simulation tests were machined in the form of a cylinder of 8.0 mm diameter and 10 mm height. The thermomechanical treatments were performed using a Gleeble 1500 thermomechanical simulator. The processing schedules employed are illustrated in Fig. 1. In the first schedule, no deformation is applied to the material; the specimen is austenitized at 860 °C for 15 min before being cooled to 300 °C, at 10 °C s^{-1}, and isothermally transformed for 90 min. The isothermal holding temperature was controlled within the range of ± 0.5 °C. After isothermal holding, the specimen was cooled to room temperature at a cooling rate of 30 °C s^{-1}. The effect of deformation was investigated at two different temperatures and strains. Schedule 1 in Fig. 1 was used to study the influence of high temperature deformation on bainitic transformation. The specimens were deformed, in compression, at 860 °C to strains of 0.25

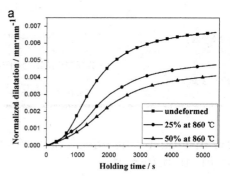

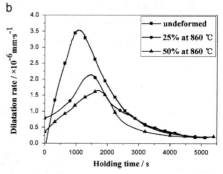

Fig. 2. (a) Dilation curves and (b) transformation rates showing the effect of ausforming at high temperature on bainitic transformation.

and 0.50. The effect of low temperature deformation was studied using schedule 2 in which the specimens were deformed at 300 °C to strains of 0.25 and 0.50 before transformation. In all cases, the diameter change during the isothermal bainitic transformation was recorded using a laser extensometer. The specimens were also examined using standard optical metallography and Scanning Electron Microscopy (SEM).

3. Results

3.1. Effect of ausforming at high temperature on bainitic transformation

Fig. 2 shows the recorded diameter changes during isothermal holding at 300 °C after ausforming at 860 °C. To a first order, the change in diameter is attributed to the volume change associated with the precipitation of bainite. Second order effects such as variant selection and transformation plasticity could occasionally complicate the interpretation of the recorded diameter changes. In the present case, however, the evolution of the diameter is largely due to the precipitation of bainite and, therefore, provides direct information on the kinetics of the reaction. In Fig. 2, the dilatation amount is normalized by dividing the instantaneous dilation by the sample diameter before transformation. It is clear from Fig. 2 that the largest volume change occurred for the non-deformed specimen. The application of a strain of 0.25 reduced the volume change obtained at the end of the transformation. Increasing the strain to 0.50 had the effect of further retarding the transformation

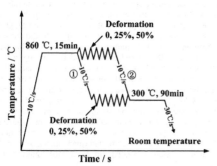

Fig. 1. Ausforming experiment procedures.

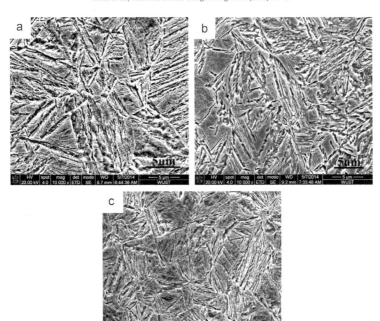

Fig. 3. SEM microstructures after bainitic transformation for 30 min at 300 °C: (a) no deformation, (b) 25% deformation at 860 °C and (c) 50% deformation at 860 °C.

and resulted in a longer incubation period and the smallest total volume change.

Not only is the maximum growth rate in the deformed material smaller, but also the maximum growth rate appears later, which means the bainite growth rate is reduced, leading to smaller terminal bainite fraction as shown in Fig. 2a. These transformation rates are consistent with the retardation of bainite transformation caused by mechanical stabilization of austenite.

The SEM micrographs showing bainite morphology in the specimens that were ausformed at 860 °C and isothermally transformed for 30 min at 300 °C are presented in Fig. 3. It is clear from these figures that the bainite sheaves become finer after deformation compared with the undeformed material. The bainite sheaves are further refined with increasing strain. This is consistent with the argument that the mechanical stabilization is caused by the accumulated debris of dislocations in the austenite which hinders growth of the bainite plates [6].

3.2. Effect of ausforming at low temperature on bainitic transformation

Fig. 4 shows the recorded diameter changes during bainite precipitation in specimens deformed and isothermally transformed at 300 °C. Interestingly, the specimen that was deformed to a strain of 0.25 appears to show a larger diameter change than the specimen without deformation. It should be pointed out that the interpretation of the results is complicated by the fact that the diameter change may be the combined result of dilatation and shape change due to variant selection or plastic flow. Gong et al. [18,19] clearly showed that ausforming at 300 °C can lead to strong variant selection of bainite. Miyamoto et al. [20] and Pereloma et al. [21] also found that ausforming favored the formation of certain bainitic variants. Because of the shear strain associated with the transformation, strong variant selection can result in non-uniform expansion during the transformation. Under these conditions, the diameter change would not exactly correspond to the dilatation. Nonetheless, qualitative results could be obtained from Fig. 4 by multiplying the normalized diameter change of the specimen deformed at 25% by a correction factor equal to the ratio of the terminal diameter change of the non-deformed specimen to that of the deformed specimen. This simple first order correction assumes that both specimens become completely bainitic at long times. The modified results are shown in Fig. 5. The results qualitatively show that a deformation of 25% accelerates the initial transformation kinetics. This conclusion is supported by optical and SEM observations. Analysis of the material which was deformed to a reduction of 50% shows that the final amount of bainite decreased compared with sample without prior deformation. It is likely that a large strain causes some mechanical stabilization of austenite. Although strong variant selection of bainite may also occur in 50% deformed sample, mechanical stabilization of austenite leads to reduction of final dilatation. Several researchers [12–17] have reported that the bainitic transformation rate is at first accelerated by deformation but eventually retarded relative to the transformation rate in the undeformed material. This is consistent with the present results after strain of 50%.

The SEM micrographs showing bainite morphology after transformation for 30 min following different strains at 300 °C are

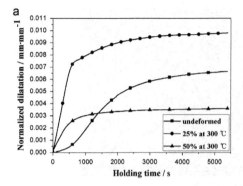

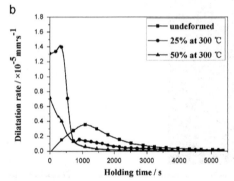

Fig. 4. (a) Dilation curves and (b) transformation rates showing the effect of ausforming at low temperature on bainitic transformation.

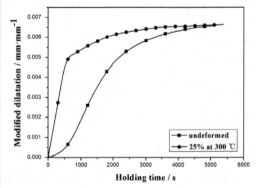

Fig. 5. Modified dilation curves showing the effect of 25% deformation at 300 °C on the bainite precipitation kinetics.

presented in Fig. 6. These micrographs demonstrate that the bainite sheaves are obviously finer after 25% deformation compared with those in the undeformed sample. Also the amount of bainite present after 30 min at 300 °C is largest for the sample with a 0.25 strain. Deformed austenite contains more deformation faults which provide more nucleation sites for the bainitic transformation, resulting in finer bainite morphology (Fig. 6b). On the contrary, when the strain increases to 50%, the austenite grains contain large amount of high dislocation density regions, lead-

ing to the mechanical stabilization of austenite. As a result the growth of bainite sheaves is significantly hindered. In addition, through the in-situ observation of bainite nucleation and growth, it was found that bainite sheaves nucleate not only on grain boundaries, but also in grains, at twin lines and pre-formed bainite plates [22–24]. The presence of more defects and dislocations after deformation provides more nucleation sites for bainitic transformation. These two factors result in the broken bainite morphology shown in Fig. 6c.

In order to investigate the mechanical stabilization of austenite in the deformed material, the amount of Retained Austenite (RA) was determined by X-Ray Diffraction (XRD) experiments conducted using X'pert Powder diffractometer with Co Kα radiation using an acceleration voltage of 40 kV, current of 150 mA and a step of 0.06°. The volume fraction of the RA is calculated according to the procedure in Reference [25]. The RA fractions are ~8% for the non-deformed material, ~26% in the material deformed to a strain of 0.25 and ~15% in the material with the strain of 0.50. XRD diffractograms showing RA fractions for the non-deformed material as well as the materials deformed to 25% and 50% are given in Fig. 7.

As for the kinetics of the transformation, small deformation strains (i.e. 0.25) at low temperature (300 °C) can lead to accelerated bainite precipitation kinetics and an increase in the final bainite volume fraction compared to the non-deformed material. The acceleration of bainite precipitation along with the increase in the amount of retained austenite implies that less martensite will be present once the ausformed ($\varepsilon=0.25$) specimens are quenched to room temperature. The amount of RA in the material deformed to 50% increased compared with sample without prior deformation, indicating that a large strain causes mechanical stabilization of austenite. At the same time sluggish ferrite growth kinetics in the specimen that was deformed 50% at 300 °C resulted in less carbon rejection into the austenite compared to the specimen that was deformed by 25% only. As a result, the amount of retained austenite in the specimen that was deformed to 50% was lower than that in the specimen deformed to 25%.

3.3. Effect of stress during bainitic transformation

Our experiments focused on the effect of pre-deformation/ausforming on the bainitic transformation. The literature clearly shows an additional effect related to the influence of an applied stress on the transformation kinetics [26–28]. Thus it is necessary to analyze the stress on samples during transformation in order to ensure that, in the present work, the stresses present during isothermal holding are not going to influence the transformation kinetics. Umemoto et al. [29] claimed that only when the stress exceeds yield strength is the transformation obviously affected by stress. A stress as small as possible is applied on samples during bainitic transformation at 300 °C in order to hold the specimen in place. Fig. 8 illustrates the stress on the samples during the bainitic transformation at different deformation conditions. The yield strength of tested steel at 300 °C is 180 MPa. It can be seen that maximum stress on samples is about 17.8 MPa, which means that the influence of stress on transformation can be ignored.

In addition, the transformation is affected by temperature. The temperature variation during isothermal holding at 300 °C was controlled within the range of ±0.5 °C. Therefore, the effect of temperature fluctuations is negligible.

4. Discussions

In present study, 25% reduction at high temperature (860 °C) retarded the bainitic transformation. Increasing the strain further

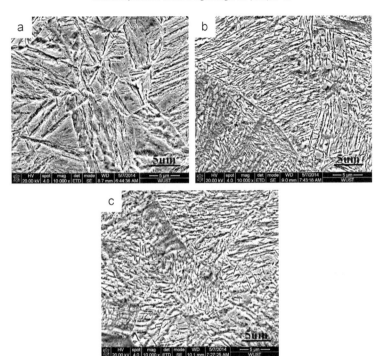

Fig. 6. SEM microstructures after bainitic transformation for 30 min at 300 °C. (a) No deformation, (b) 25% deformation at 300 °C and (c) 50% deformation at 300 °C.

resulted in additional retardation of the transformation. Most researches show that prior deformation of austenite hinders the following bainitic transition [7,9] due to the mechanical stabilization of austenite caused by ausforming. Bhadeshia et al. [12,29] explained mechanical stabilization in terms of the structure of the transformation interface. Displacive transformations occur by the advance of glissile interfaces which can be rendered sessile when they encounter dislocation debris. Thus deformation can retard the decomposition of austenite. Most of the work on mechanical stabilization effects has focused on the martensitic transformation with few studies on bainite. Several researchers have demonstrated that the growth of bainite is retarded by the deformation debris in the austenite [6,12,13].

The bainite transformation was also hindered by 50% reduction at 300 °C. Several research works have been reported to explain the retardation of bainite precipitation by ausforming. Some researchers [6,13,30] suggested that while heterogeneous nucleation becomes more frequent as defects are introduced into the austenite, growth by a displacive mechanism is stifled as the interface encounters forests of dislocations. Deformed austenite therefore transforms to a smaller quantity of bainite than undeformed austenite, and any bainite that forms is more refined. Shipway et al. [6] claimed that once nucleated, the bainite is prevented from growing by the dislocation debris in the austenite, and the amount of transformation per plate of bainite is reduced. Singh and Bhadeshia [12,13,31] emphasized the fact that deformation is not uniform in the sample. The lightly deformed regions transform more rapidly relative to undeformed austenite because of the increase in the number of potential nucleation site, leading to accelerated transformation kinetics. Although the nucleation rate is larger in the heavily deformed regions, the overall rate of transformation is reduced because each nucleus then transforms to a smaller volume due to mechanical stabilization of the interface, resulting in the smaller terminal fraction of bainite. In present work, the initial transformation rate of bainite after 50% deformation at 300 °C is larger than undeformed austenite due to the increased number of nucleation sites (Fig. 4b). Nevertheless the transformation rate after 50% deformation decreases continually during the whole transformation process (Fig. 4b), indicating that the growth of bainite is retarded from the beginning to the end of transformation. The results are consistent with the existing explanations for the retardation of bainite precipitation as a result of prior-deformation/ausforming.

It is very interesting to note that a prior deformation of 25% at 300 °C accelerates the initial transformation rate. Compared with the undeformed sample, the deformed sample with small strain has more defects, providing more nucleation sites for bainitic transformation. Meanwhile, dislocation debris caused by deformation hinders the growth of bainite sheaves. However, the retardation due to the dislocation debris is small in this case because of small deformation applied to the specimen. As a result, the transformation rate is accelerated at short times. The increase of nucleation sites plays a more important role in the transformation than the retardation of growth at the beginning, leading to a larger bainite fraction at the early stage.

The results in present study suggest that the rate of transformation changes non-monotonically with the ausforming strain. It is, therefore, possible to identify the level of deformation that would lead to the highest transformation rate in order to reduce the processing time of carbide-free bainitic steels. The presence of

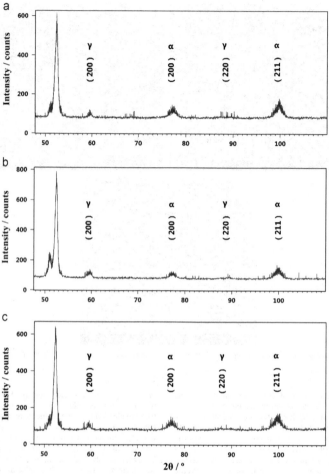

Fig. 7. XRD diffractograms showing RA fractions. (a) No deformation, (b) 25% deformation at 300 °C and (c) 50% deformation at 300 °C.

an increased volume fraction of retained austenite and the decrease in the amount of martensite are additional advantages for the application of small ausforming strains at low temperature. The present findings are consistent with the predictions by Chatterjee et al. [32] who proposed that the retardation of bainite precipitation occurs above a critical strain. The present study shows, however, that temperature plays a critical role. The critical strain was observed for deformation at 300 °C, but it was not observed for deformation at 860 °C. It appears that at high temperature, deformation will always have the effect of reducing the transformation kinetics. The origin of the difference between the high and low temperature behavior is no immediately obvious. One may speculate that the difference in behavior is related to recovery. Recovery after high temperature deformation would replace the random dislocation structure with a well-developed cell structure. It may be that this eliminates the additional heterogenous nucleation sites for bainite, but leaves behind misorientation changes (e.g. cell and subgrain boundaries) that arrest the growth of bainite.

5. Conclusions

The effects of prior deformation conditions on bainitic transformation were investigated by the combination of metallography and dilatometry. The influences of both deformation temperature and strain on bainitic transformation in a bainite steel were discussed. The results show that the effect of prior deformation on bainitic transformation depends on deformation temperature and strain.

The transformation is retarded by austenite deformation at high temperature. Meanwhile, the final bainite fraction decreases after deformation with larger strain at 300 °C although the initial transformation rate is accelerated. The bainitic transformation can be promoted by a small strain at 300 °C. This provides an important opportunity to reduce the processing times for carbide-free bainitic steels. In addition, there exists a critical strain at low temperature, below which the amount of bainite+austenite in the final microstructure is increased and above which the amount of bainite+austenite is decreased. From a practical point of view, this is a useful strain for the design of an industrial process.

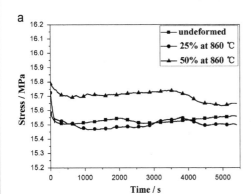

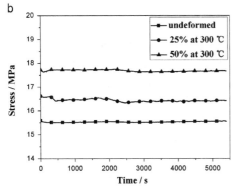

Fig. 8. Stress on samples during transformation at 300 °C. (a) High temperature deformation and (b) low temperature deformation.

Acknowledgments

HSZ gratefully acknowledges the financial support of the Natural Science and Engineering Research Council of Canada (CRDPJ424277-11). GX is grateful to the financial supports from National Natural Science Foundation of China (NSFC) (No. 51274154), the National High Technology Research and Development Program of China (No. 2012AA03A504) and the State Key Laboratory of Development and Application Technology of Automotive Steels (Baosteel Group) (BS2012-01).

References

[1] C.K. Yao, Y.M. Zhang, X.Y. Men, S.Q. Zhang, Mater. Sci. Eng. 83 (1986) 1–6.
[2] V.M. Khlestov, E.V. Konopleva, H.J. Mcqueen, Can. Metall. Q. 37 (1998) 75–79.
[3] J.R. Strife, M.J. Carr, G.S. Ansell, Metall. Trans. A 8 (1976) 1471–1484.
[4] V. Ranhavan, ASM Int. (1992) 197–226.
[5] K. Tsuzaki, S. Fukasaku, Y. Tomota, T. Maki, Mater. Trans. JIM 32 (1991) 222–228.
[6] P.H. Shipway, H.K.D.H. Bhadeshia, Mater. Sci. Technol. 11 (1995) 1116–1128.
[7] R.H. Larn, J.R. Yang, Mater. Sci. Eng. A 278 (2000) 278–291.
[8] C.S. Chiou, J.R. Yang, C.Y. Huang, Mater. Chem. Phys. 69 (2001) 113–124.
[9] J.R. Yang, C.Y. Huang, W.H. Hseich, C.S. Chiou, Mater. Trans. JIM 37 (1996) 579–585.
[10] A.T. Davenport, TMS-AIME, New York, USA, 1977, pp. 517–536.
[11] C.H. Lee, H.K.D.H. Bhadeshia, H.C. Lee, Mater. Sci. Eng. A 360 (2003) 249–257.
[12] H.K.D.H. Bhadeshia, Mater. Sci. Eng. A 273–275 (1999) 58–66.
[13] S.B. Singh, H.K.D.H. Bhadeshia, Mater. Sci. Technol. 12 (1996) 610–612.
[14] T. Maki, in: K.A. Taylor, et al. (Eds.), The Minerals, Metals and Materials Society, Warrendale, USA, 1993, pp. 3–16.
[15] R. Freiwillig, J. Kudrman, P. Chraska, Metall. Trans. A 7A (1976) 1091–1097.
[16] X.J. Jin, N. Min, K.Y. Zheng, T.Y. Hsu, Z.Y. Xu, Mater. Sci. Eng. A 438–440 (2006) 170–172.
[17] R.H. Edwards, N.F. Kennon, Metall. Trans. A 9A (1978) 1801–1809.
[18] W. Gong, Y. Tomota, M.S. Koo, Y. Adachi, Scr. Mater. 63 (2010) 819–822.
[19] W. Gong, Y. Tomota, Y. Adachi, A.M. Paradowska, J.F. Kelleher, S.Y. Zhang, Acta Mater. 61 (2013) 4142–4154.
[20] G. Miyamoto, N. Iwata, N. Takayama, T. Furuhara, J. Alloy. Compd. 577S (2013) S528–S532.
[21] E.V. Pereloma, F. Al-Harbi, A.A. Gazder, J. Alloy. Compd. 615 (2014) 96–110.
[22] Z.W. Hu, G. Xu, H.J. Hu, L. Wang, Z.L. Xue, Int. J. Miner. Metall. Mater. 21 (2014) 371–378.
[23] G. Xu, F. Liu, L. Wang, H.J. Hu, Scr. Mater. 68 (2013) 833–836.
[24] H.J. Hu, G. Xu, F. Liu, L. Wang, L.X. Zhou, Z.L. Xue, Int. J. Mater. Res. 105 (2014) 337–341.
[25] A.K. De, D.C. Murdock, M.C. Mataya, J.G. Speer, D.K. Matlock, Scr. Mater. 50 (2004) 1445–1449.
[26] H.K.D.H. Bhadeshia, S.A. David, J.M. Vitek, R.W. Read, Mater. Sci. Technol. 7 (1991) 686–698.
[27] A. Matsuzaki, H.K.D.H. Bhadeshia, H. Harade, Acta Metall. Mater. 42 (1994) 1081–1090.
[28] P.H. Shipway, H.K.D.H. Bhadeshia, Mater. Sci. Eng. A 201 (1995) 143–149.
[29] M. Umemoto, S. Bando, I. Tamura, in: I. Tamura, et al. (Ed.), Proceedings of the International Conference on Martensitic Transformation 1986, Sendai, Japan (The Jpn Inst Met, 1987:595).
[30] H.K.D.H. Bhadeshia, Bainite in Steels, 2nd ed., The Cambridge University Press, Cambridge (2001) 207.
[31] S.B. Singh, Ph.D. thesis, University of Cambridge, 1998.
[32] S. Chatterjee, H.S. Wang, J.R. Yang, H.K.D.H. Bhadeshia, Mater. Sci. Technol. 22 (2006) 641–644.

References

[1] Roberts W L. Cold Rolling of Steel [M]. New York: CRC Press, 1978.
[2] Roberts W L. Hot Rolling of Steel [M]. New York: CRC Press, 1983.
[3] Ginzburg V B. Steel-Rolling Tecnology-Theory and Practice [M]. New York: CRC Press, 1989.
[4] Richad Wechsler. The Status of Twin-roll Casting Technology [J]. Scandinavian Journal of Metallurgy, 2003, 32: 58~63.
[5] Byrer T G, et al. Forging Handbook [M]. Cleveland: Forging Industry Association, 1985.
[6] Jenson J E. Forging Industry Handbook [M]. Cleveland: Forging Industry Association, 1970.
[7] Boyer H E. Metals Handbook [M]. Cleveland: American Society for Metals, 1976.
[8] Open Die Forging Institute. Open Die Forging Manual [M]. Cleveland: Forging Industry Association, 1982.
[9] Eary D F, Reed E A. Techniques of Pressworking Sheet Metal [J]. New Jersey: Prentice Hall, 1958.
[10] Laue K, Stenger H. Extrusion: Processes, Machinery, Tooling [M]. Cleveland: American Society for Metals, 1981.
[11] Müller E. Hydraulic Extrusion Presses [M]. Berlin: Springer, 1961.
[12] Gourd L M. Principles of Welding Technology [M]. London: Edward Arnold, 1980.
[13] Charlotte, Weisman. Welding Handbook [M]. Miami: American Welding Society, 1976.
[14] Houldcroft P T. Welding Process Technology [M]. Cambridge: Cambridge University Press, 1977.
[15] AWS Committee. Definitions, Symbols and Metric Practice under the Direction of AWS Technical [M]. Miami: American Welding Society, 1976.
[16] Klas Weman. Welding Process Handbook [M]. Amsterdam: Elsevier, 2003.
[17] Cary H B. Modern Welding Technology [M]. New Jersey: Prentice Hall, 2002.
[18] 《铸造专业英语文选》编写组. 铸造专业英语文选 [M]. 南宁: 广西人民出版社, 1979.
[19] Rao P N. 制造技术: 铸造、成形和焊接 [M]. 北京: 机械工业出版社, 2003.
[20] 张军. 材料专业英语译写教程 [M]. 北京: 机械工业出版社, 2001.
[21] Callister W D, Rethwisch D G. Materials Science and Engineering: An Introduction [M]. New Jersey: Wiley & Sons, 2007.
[22] Xu G, Liu F, Wang L, et al. A New Approach to Quantitative Analysis of Bainitic Transformation in a Superbainite Steel [J]. Scripta Materialia, 2013, 68 (11): 833~836.
[23] Xu G, Gan X L, Ma G J, et al. The Development of Ti-alloyed High Strength Microalloy Steel [J]. Materials and Design, 2010, 31 (6): 2891~2896.
[24] Hu H J, Zurob H S, Xu G, et al. New Insights to the Effects of Ausforming on the Bainitic Transformation [J]. Materials Science and Engineering A, 2015, 626: 34~40.
[25] Hu H J, Xu G, Wang L, et al. The Effects of Nb and Mo Addition on Transformation and Properties in Low Carbon Bainitic Steels [J]. Materials and Design, 2015, 84 (4): 95~99.